THE WASTE PRODUCTS OF AGRICULTURE

Their Utilization as Humus

by

ALBERT HOWARD, C.I.E., M.A.
Director of the Institute Of Plant Industry, Indore, and Agricultural
Adviser to States in Central India and Rajputana

and

YESHWANT D. WAD, M. Sc.

Chief Assistant in Chemistry, Institute of Plant Industry, Indore

To

SIR REGINALD GLANCY K.C.I.E., C.S.I., C.I.E., I.C.S.
Member of the Council of India Formerly Agent to the Governor-
General in Central India First President of the Board of Governors of
the Institute of Plant Industry, Indore (1924-1929)

Of composts shall the Muse disdain to sing?
Nor soil her heavenly plumes? The sacred Muse
Nought sordid deems, but what is base nought fair,
Unless true Virtue stamp it with her seal.
Then, planter, wouldst thou double thine estate,
Never, ah! never, be asham'd to tread
Thy dung-heaps.

(From Grainger's The Sugar Cane.)

Contents

PREFACE

One of the main features of crop production at the present day is waste. Except in the Far East, where the large indigenous population has to be fed from the produce of the country-side, little is being done to utilize completely the by-products of the farm in maintaining the fertility of the soil. The ever-growing supplies of agricultural produce, needed by industry and trade, have been provided either by taking up new land or by the purchase of artificial manures. Both these methods are uneconomic. The exploitation of virgin soil is a form of plunder. Any expenditure on fertilizers which can be avoided raises the cost of production, and therefore reduces the margin of profit. It needs no argument to urge that, in maintaining the fertility of the soil, the most careful attention should be paid to the utilization of the waste products of agriculture itself before any demands are made on capital--natural or acquired.

For the last twenty-six years, the senior author has been engaged in the study of crop production in India and in devising means by which the produce of the soil could be increased by methods within the resources of the small holder. These investigations fell into two divisions: (1) the improvement of the variety; and (2) the intensive cultivation of the new types. In the work of replacing the indigenous crops of India by higher yielding varieties, it was soon realized that the full possibilities in plant breeding could only be achieved when the soil in which the improved types are grown is provided with an adequate supply of organic matter in the right condition. Improved varieties by themselves could be relied on to give an increased yield in the neighbourhood of ten per cent. Improved varieties plus better soil conditions were found to produce an increment up to a hundred per cent or even more.

THE WASTE PRODUCTS OF AGRICULTURE

Steps were therefore taken: (1) to study the conversion of all forms of vegetable and animal wastes into organic matter (humus) suitable for the needs of the growing crop; and (2) to work out a simple process by which the Indian cultivator could prepare an adequate supply of this material from the by-products of his holding. In other words he has been shown how to become a chemical manufacturer. This task involved a careful study of the various systems of agriculture which so far have been evolved and particularly of the methods by which they replenish the soil organic matter. The line of advance in raising crop production in India to a much higher level then became clear. Very marked progress could be made by welding the various fragments of this subject--the care of the manure heap, green-manuring and the preparation of artificial farmyard manure--into a single process, which could be worked continuously throughout the year and which could be relied upon to yield a supply of humus, uniform in chemical composition and ready for incorporation into the soil. This has been accomplished at the Institute of Plant Industry at Indore. The work is now being taken up in Sind and at various centres in Central India and Rajputana.

The Indore process for the manufacture of humus is described in detail in the following pages. It can be adopted as it stands throughout the tropics and sub-tropics, and also on the small holdings and allotments of the temperate zone. How rapidly the method can be incorporated into the large-scale agriculture of the west is a question which experience alone can answer. It will in all probability depend on how far the process can be mechanized.

In the field of rural hygiene there is great scope for the new method. It can be applied to the utilization of all human, animal and vegetable wastes in such a manner that the breeding of flies is prevented, the water and the food-supply of the people safeguarded and the general health of the locality improved. Cleaner and healthier villages will then go hand in hand with heavier crops.

A.H. Y.D.W. Indore 6 April, 1931

I. INTRODUCTION

The maintenance of the fertility of the soil is the first condition of any permanent system of agriculture. In the ordinary processes of crop production, fertility is steadily lost; its continuous restoration by means of manuring and soil management is therefore imperative.

In considering how the ideal method of manuring and of soil management can be devised, the first step is to bring under review the various systems of agriculture which so far have been evolved. These fall for the most part into two main groups: (1) the methods of the Occident to which a large amount of scientific attention has been devoted during the last fifty years; and (2) the practices of the Orient which have been almost unaffected by western science.

The systems of agriculture of the Occident and of the Orient will now be briefly considered with a view of extracting from each ideas and results which can be utilized in the evolution of the ideal method of maintaining and increasing the fertility of the soil.

NOTE: In the general organization of agriculture, Europe stands mid-way between the east and the west and provides, as it were, the connecting link between these two methods of farming.

THE AGRICULTURAL SYSTEMS OF THE OCCIDENT

The most striking characteristic of the agriculture of the west is the comparatively large size of the holding. Large farms are the rule; small holdings are the exception.

THE WASTE PRODUCTS OF AGRICULTURE

NOTE: The growth of allotments for the production of vegetables in the neighbourhood of urban areas is a comparatively recent phenomenon and only affects a small area.

The large farms of the west are for the most part engaged in the production of food and a few raw materials like wool for the urban populations of the world, which are mainly concerned with manufacture and trade. To produce these vast supplies, and at the same time to place them on the markets at low rates, practically all the unoccupied temperate regions of the world, which are suitable for the white races, have already been utilized. The best areas of North America, of the Argentine, of South Africa and large tracts of Australia and practically the whole of New Zealand have during the last hundred years been exploited to produce the endless procession of cargoes of food and raw materials required by the great markets of the world.

The weakness of this system of agriculture lies in the fact that it is new and has not yet received the support which centuries of successful experience alone can provide. At first it was based on the exploitation of the stores of organic matter accumulated by virgin land, which at the best could not last for more than a limited number of years. Even now there is practically no attempt to utilize the large quantities of wheat straw and other vegetable wastes for keeping up the store of organic matter in the soil. The new areas of North America for example soon showed signs of exhaustion. Manuring has become necessary as in the case of the older fields of Europe. To supply the large quantities of combined nitrogen needed, all possible sources except the right one--the systematic conversion of the waste products of agriculture into humus--have one after the other been utilized: guano from the islands off the Peruvian coast, nitrate of soda from Chile, sulphate of ammonia from coal and more recently synthetic nitrogen compounds obtained from the atmosphere. These substances are supplemented by another class of nitrogenous organic manures such as artificial guanos, dried blood and slaughter-house residues, oil cakes and wool waste--the by-products of agriculture--and by another group of artificials--the various phosphatic and potassic fertilizers. These supplies of concentrated manures have enabled agricultural production to be kept at a high level. The fact of their existence for a time tended to distract attention from the fullest utilization of the by-products of the farm. Recently, however, a change has taken place and a large amount of scientific effort has been devoted to the problems which

4

centre round the waste products, both animal and vegetable, of agriculture itself. The need of keeping up the supply of organic matter in the soil is now widely recognized.

After the large size of the holding and the necessity of manuring, the high cost of labour is another leading characteristic of western farming. The number of men per square mile of agricultural land who actually work is low.

NOTE: The comparative figures of crop production per worker for the five-year period preceding the War, prepared by the United States Department of Agriculture, are instructive. The number of workers employed per 1,000 acres of crop land was approximately 235 in Italy, 160 in Germany, 120 in France, 105 in England and Wales, 60 in Scotland but only 41 in the United States. In Canada, according to Riddell, the 1911 figures show that every 1,000 acres called for only 26 workers. This observer states that in the three prairie provinces (Alberta, Manitoba, Saskatchewan) the figures are even more striking: the area under field crops was 17,677,091 acres, and the numbers engaged in agriculture was 283,472, so that each person so employed was responsible for 62 acres. Every 1,000 acres required only 16 workers. Since these data were published, further statements have appeared from which it would seem that the size of the working population in agriculture in North America has shrunk still further.

This state of things has arisen from the dearness and scarcity of labour, which has naturally led to the study of labour-saving devices including the use of machinery. Whenever a machine can be invented which saves human labour its spread is rapid. Engines of various kinds are the rule everywhere. The inevitable march of the combine-harvester, in all the wheat-producing areas of the world, is the latest example of the mechanization of the agriculture of the west. Another feature of this extensive system of large-scale agriculture is the development of food preservation processes, of transport and of marketing, by which the products of agriculture are cheaply and rapidly moved from the field to the centres of distribution and consumption. There is no great dearth of capital at any stage. Money can always be found for any new machine and for any new development which is likely to return a dividend. Land and capital are abundant; efficient transport and good markets abound. The comparatively small supply of suitable labour and its high cost provide the chief agricultural problems of the west.

This system of agriculture is essentially modern and has developed largely as one of the consequences of the discovery of the steam engine and the rapid exploitation of the supplies of coal, oil and water-power. It has only been made possible by the existence of vast areas of virgin land in parts of the earth's surface on which the white races can live and work. As already mentioned the weak point in this method of crop production is that it is new and lacks the backing which only a long period of practical experience can supply. Mother Earth is provided with an abundant store of reserve fertility which can always be exploited for a time. Every really successful system of agriculture however must be based on the long view, otherwise the day of reckoning is certain.

Side by side with this method of utilizing the land there has been a great development of science. Efforts have been made to enlist the help of a number of separate sciences in studying the problems of agriculture and in increasing the production of the soil. This has entailed the foundation of numerous experiment stations, which every year pour out a large volume of printed results and advice to the farmer. At first the scientific workers naturally devoted themselves to solving local problems and to furnishing scientific explanations of various agricultural practices. This phase is now passing. A new note is beginning to appear in the publications of the experiment stations, namely that of direction and advice which can only be advanced by men whose education and training combine the ideas of science with the aims of the statesman. The feeling is not only growing but is being expressed that it is no longer the business of science merely to solve the problems of the moment. Something more is needed. The chief function of science in the agriculture of the future is to provide intelligent direction in general policy and to point the way.

THE WASTE PRODUCTS OF AGRICULTURE

THE AGRICULTURAL SYSTEMS OF THE ORIENT

Peasant Holdings

The chief feature of the agricultural systems of the east is the small size of the holding. The relation between man-power and cultivated area in India is given in Table I. In this table, based on the Census Report of 1921, the number of workers and the acreage cultivated have been calculated for the chief provinces of British India. Incidentally these figures illustrate how intense is the struggle for existence in this portion of the tropics.

TABLE I.--THE RELATION BETWEEN MAN-POWER AND CULTIVATED AREA IN INDIA

Provinces	Number of acres cultivated by 100 ordinary cultivators
Bombay	1,215
North-West Frontier Province	1,122
Punjab	918
Central Provinces	848
Burma	565
Madras	491
Bengal	312
Bihar and Orissa	309
Assam	296
United Provinces	251

These minute holdings are frequently cultivated by extensive methods (those suitable for large areas) which neither utilize the full energies of man and beast nor the potential fertility of the soil. Such a system of agriculture can only result in poverty. The obvious line of advance is the gradual introduction of more intensive methods, for which the supply of suitable manure, within the means of the average cultivator, is bound to prove an important factor.

If we turn to the Far East, to China and Japan, a similar system of small holdings is accompanied by an even more intense pressure of

7

population both human and bovine. In the introduction to Farmers of Forty Centuries, King states that the three main islands of Japan had in 1907 a population of 46,977,003, maintained on 20,000 square miles of cultivated fields. This is at the rate of 2,349 to the square mile or more than three people to each acre. Note that these figures agree very closely with those quoted in the Japan Year Book of 1931 in which the number of persons per square kilometre is given as 969: equivalent to 2,433 to the square mile.

In addition Japan fed on each square mile of cultivation a very large animal population--69 horses and 56 cattle, nearly all employed in labour; 825 poultry; 13 swine, goats and sheep. Although no accurate statistics are available in China, the examples quoted by King reveal a condition of affairs not unlike Japan. In the Shantung Province, a farmer with a family of twelve kept one donkey, one cow and two pigs on 2.5 acres of cultivated land--a density of population at the rate of 3,072 people, 256 donkeys, 256 cattle and 512 pigs per square mile. The average of seven Chinese holdings visited gave a maintenance capacity of 1,783 people, 212 cattle or donkeys and 399 pigs--nearly 2,000 consumers and 400 rough food transformers per square mile of farm land. In comparison with these remarkable figures, the corresponding statistics for 1900 in the case of the United States per square mile were: population 61, horses and mules 30.

The problems of tropical agriculture for the most part relate to small holdings. The main purpose of this peasant agriculture is crop production; animal husbandry is much less important. In India the crops grown fall into two classes--(1) food and fodder crops and (2) money crops. The former includes, in order of area: rice, millets, wheat, pulses and fodder crops, barley and maize and sugar-cane. The money crops are more varied; cotton and oil seeds are the most important, followed by jute and other fibres, tobacco, tea, opium, indigo and coffee. It will be seen that food and fodder crops comprise 82 per cent of the total area under crops and that money crops, as far as extent is concerned, are relatively unimportant.

THE WASTE PRODUCTS OF AGRICULTURE

TABLE II.--AGRICULTURAL STATISTICS OF BRITISH INDIA, 1926-27
Area, in acres, under food and fodder crops

Rice	78,502,000
Millets	38,776,000
Wheat	24,181,000
Gram	14,664,000
Pulses and other good grains	29,154,000
Fodder crops	8,940,000
Condiments, spices, fruits, vegetables, and misc. food crops	7,537,000
Barley	6,387,000
Maize	5,555,000
Sugar	3,041,000
TOTAL, FOOD & FODDER CROPS	216,737,000

Area, in acres, under money crops

Cotton	15,687,000
Oil seeds, chiefly rape and mustard, sesamum, groundnuts and linseed	14,999,000
Jute and other fibres	4,411,000
Dyes, tanning materials, drugs, narcotics and miscellaneous crops	1,729,000
Tobacco	1,055,000
Tea	738,000
Opium	59,000
Indigo	104,000
Coffee	91,000
TOTAL, MONEY CROPS	38,873,000

The primary function of Indian agriculture is to supply the cultivator and his cattle with food. Compared with this duty all other matters are subsidiary. The houses are built of mud, thatched with grass and are almost devoid of furniture. Expenditure on clothing and warmth is, on account of the customs of the country and the nature of the climate, much smaller than in European countries. Nevertheless, the cultivators require a little money with which to pay the land revenue and to purchase a few necessaries in the village markets. Hence the growth of money crops to the extent of about one-fifth the total cultivated area. The produce, after conversion into cash, is afterwards either worked up in the local mills or exported. To some extent food crops are also money crops. The population of the towns and cities is largely fed from the produce of the soil, while in addition a small percentage of the total food grains produced is exported to foreign countries. In some crops like sugar-cane, the total out-turn is insufficient for the towns and large quantities of sugar are imported from Java, Mauritius and the continent of Europe.

When we come to the details of soil management, a further striking difference between the methods in vogue in the west and on the peasant holdings of the east is at once manifest. In China, fertility has for centuries been maintained at a high level without the importation of artificial manures. Although it was not till 1888, after a protracted controversy lasting thirty years, that western science finally accepted as proved the important part played by pulse crops in enriching the soil, nevertheless centuries of experience had taught the peasants of the east the same lesson. The leguminous crop in the rotation is everywhere one of their old fixed practices. Moreover, on the alluvium of the Indo-Gangetic Plain, the deep, spreading root-system of the pigeon pea (Cajanus indicus Spreng.) is utilized by the peasantry as an efficient substitute for the periodical subsoil ploughing which these closely-packed, silt-like soils require. In the case of the best cultivators, the general soil management and particularly the conservation and utilization of combined nitrogen has already reached a high level. This has been described, in the case of the United Provinces of India, by Clarke in a recent paper which has been reproduced as Appendix B. In China and Japan not only the method of soil management but also the great attention that is paid to the systematic preparation, outside the field, of food materials for the crop from all kinds of vegetable and animal wastes compelled the admiration of one of the most brilliant of the agricultural investigators of the last generation. The results are set out by King in his unfinished work--Farmers of Forty Centuries--which should be prescribed as a textbook in every agricultural school and college in the world.

Another feature of this agriculture is the cultivation of rice wherever the soil and water-supply permit. In the scientific consideration of the methods of soil management under which the rice crop of the Orient is produced, practical experience at first seems to contradict one of the great principles of the agricultural science of the Occident, namely the dependence of cereals on nitrogenous manures. Large crops of rice are produced in many parts of India on the same land year after year without the addition of any manure whatever. The rice fields of the country export paddy in large quantities to the centres of population or abroad, but there is no corresponding import of combined nitrogen.

Taking Burma as an example of an area exporting rice beyond seas, during the twenty years ending 1924, about 25,000,000 tons of paddy have been exported from a tract roughly 10,000,000 acres in area. As unhusked rice contains about 1.2 per cent of nitrogen the amount of this element, shipped overseas during twenty years or destroyed in the burning of the husk, is in the neighbourhood of 300,000 tons. As this constant drain of nitrogen is not made up for by the import of manure, we should expect to find a gradual loss of fertility. Nevertheless this does not take place either in Burma or in Bengal, where rice has been grown on the same land year after year for centuries. Nearly the soil must obtain fresh supplies of nitrogen from somewhere, otherwise the crop would cease to grow. The only likely source is fixation from the atmosphere, probably in the submerged algal film on the surface of the mud. This is one of the problems of tropical agriculture which calls for early investigation.

Another important difference between the east and the west concerns the supply of labour. In the Orient it is everywhere adequate, as would naturally follow from the great density of the rural population. Indeed in India it is so abundant that if the time wasted by the cultivators and their cattle for a single year could be calculated as money, at the local rates of labour, a perfectly colossal figure would be obtained. One of the problems underlying the development of agriculture in India is the discovery of the best means of utilizing this constant drain, in the shape of wasted hours, for increasing crop production. There is therefore no lack of human labour in developing the agriculture of the east. Another favourable factor is the existence of excellent breeds of work-cattle and of the buffalo.

THE WASTE PRODUCTS OF AGRICULTURE

NOTE: The buffalo is the milk cow of the Orient and is capable not only of useful labour in the cultivation of rice, but also of living and producing large quantities of rich milk on a diet on which the best dairy cows of Europe and America would starve. The digestive processes of the buffalo is a subject which appears to have escaped the attention of the investigators of animal nutrition.

The last characteristic of this ancient system of agriculture is lack of money. Again there is a great contrast between the east and the west. There is little or no spare capital for the improvement of the holding. Over large tracts of India at any rate, the cultivators are in the hands of the moneylender and indebtedness is the rule. For many years one of the pre-occupations of Government has been the discovery of safeguards by which the cultivator can be saved from the worst consequences of his own folly--reckless borrowing for unproductive purposes--and maintained on the land. The recent development of co-operation and the rapid increase in the number of primary credit societies has only been possible because of this volume of indebtedness.

PLANTATIONS

While small holdings, accompanied by a dense population, are an important feature of eastern agriculture, nevertheless there are exceptions. Throughout this portion of the tropics European enterprise has removed the original forest and established in its place extensive plantations of such crops as sugar-cane, tea, rubber and coffee. The labour for these estates is obtained from indigenous sources; the capital and management are contributed by Europeans. Plantations of this kind are common all over the east and are an important feature of the agriculture of Java, Ceylon, the Federated Malay States, Assam and the uplands of Southern India. One of the features of this agriculture is the attention paid to manurial problems. Comparatively large sums of money are expended every year in the purchase of artificial manures, mainly for keeping up the supply of combined nitrogen. During a tour in Ceylon in 1908, when visits were paid by the senior author to a number of tea estates, the managers invariably produced their manurial programme on which suggestions were always invited. Ceylon at that time offered a tragic example of the damage which results from uncontrolled tropical rainfall on sloping land, from which the forest canopy had been removed without providing a proper system of terracing combined with surface-drainage. Over large areas of hilly country, formerly forest and now exclusively under tea, practically the whole of the valuable surface soil rich in humus had been lost by denudation. The tea plant was producing crops from the relatively poor subsoil, supplemented by the constant application of expensive manures.

In a recent review of this question in Crop Production in India published in 1924, the damage which has resulted from erosion on the plantations of the Orient was referred to (pp. 14-5) as follows:--

It is in the planting areas of the east, however, that the most striking examples of soil denudation are to be found. Instances of damage to the natural capital of the country are to be seen on the tea estates near Darjeeling, on the hill-sides in Sikkim on the upper terraces in the vale of Kashmir, in the Kumaon Hills, on the tea estates in Ceylon and Assam, and in the planting districts of Southern India and the Federated Malay States. In most of these areas forest land was so abundant that the need for the preservation of the soil was not at first recognized. Thanks to the efforts of Hope, a former scientific officer employed by the tea industry in Assam the control of the drainage and the checking of erosion are now widely recognized and are being dealt with by the planters in many parts of India. A great impetus to this work was given by the publication in India of a detailed account of the methods in use by the Dutch planters in Java, where the terracing and drainage of sloping land, under tea and other crops has been carried to a high stage of perfection. In this island the area of land available for planting is strictly limited, while the feeding of the large indigenous population is always a serious problem. As a consequence the development of the island is very strictly controlled by the Government, and one of the conditions of planting new forest lands is the provision of a suitable system of terraces combined with surface-drainage. The advantage is not all on the side of the State. The manuring of tea soils in Java is far less necessary than in Ceylon and India, while one important consequence of the retention of the valuable soil made by the forest is healthy growth, which suffers remarkably little damage from insect and fungoid pests.

THE WASTE PRODUCTS OF AGRICULTURE

UNDEVELOPED AREAS

Very large stretches of the Orient are still under forest and at present carry a very small population, supported by hunting, fishing and by the small cultivated areas surrounding the villages. These undeveloped forest areas occur everywhere, particularly in the Malay Archipelago, the Federated Malay States, Burma and the low country of Ceylon. In the search for the ideal method of manuring in the tropics, the greatest care will have to be taken to preserve the valuable surface soil whenever the forest canopy has to be removed for the creation of new cultivated land. Some at any rate of these potentially rich tracts are almost certain to be taken up during the present century. They will therefore provide ample opportunities of applying any lessons in soil management, which science can extract from experiment and from experience. The serious mistakes of the past must not be repeated when the time comes for developing the vast areas of tropical forest still untouched.

It will be evident that the systems of agriculture of the west and of the east are very different and that the two have little or nothing in common. In a sense these two methods of managing land remind one of the two sides of a coin. The one supplements the other: each can be regarded as a part of one great whole. Clearly when attempting to evolve the ideal system of manuring and soil management of the future, both of these widely different methods of agriculture must be studied. This has been done by the senior author for the last twenty-six years in various parts of India--on the alluvium of the Indo-Gangetic plain at Pusa in Bihar, on the loess soils of the Quetta Valley on the Western Frontier and on the black cotton soils of peninsular India at Indore. The chief climatic factors at Indore are represented in Plate II. The climate of Quetta resembles generally that of Persia, where the rainfall is received mainly during the winter months, often in the form of snow. At these three centres a method of utilizing all the vegetable and animal wastes of the holding has gradually been evolved. The latest scientific work of the Occident and particularly that recently accomplished at the experiment station of New Jersey, together with the practices in vogue in India and the Far East, have been welded together and synthesized into a system for the continuous manufacture of manure throughout the year so that it forms an integral part of the industry of agriculture.

THE WASTE PRODUCTS OF AGRICULTURE

In considering all this information--the various agricultural systems in use at the present time, as well as the large volume of scientific papers dealing with manurial questions, which have been poured out by the experiment stations during the last fifty years, we have been impressed by the evils inseparable from the present fragmentation of any large agricultural problem and its attack by way of the separate science. All this seems to follow from the excessive specialization which is now taking place, both in the teaching and in the application of science. In the training given to the students and in much of the published work, the tendency of knowing more and more about less and less is every year becoming more marked. For this reason any review of the problem of increasing soil fertility is rendered peculiarly difficult, not only by the vast mass of published papers but also by their fragmentary and piecemeal nature.

No extra labour is required in our manure factory. No imported chemicals such as Adco are needed in this process. No capital is required at any stage of the manufacture. The methods now in use at Indore form the main subject of this book, which also attempts to deal with a number of related matters such as--the role of organic matter in the soil, the methods of replenishing the supply of organic matter now in use and the recent investigations which have been carried out on the conditions necessary for converting raw organic residues into humus which can be immediately nitrified in the soil and so made use of by the plant. The Indore process can easily be carried out, not only in the tropics but also on the small holdings of the temperate regions and on the allotments (provided space is made available) in the neighbourhood of urban areas, where it is now the practice to burn most of the vegetable waste. How rapidly the system can be introduced into the farming systems of the Occident is a question to which no answer can be given until the ideas in this book have been fully tried out in western agriculture. It is not impossible that they may founder for a time on the present high cost of labour. The method however is in full accord with the well-marked tendency in western agriculture towards a more intensive production. The inevitable change over from extensive to intensive methods has already begun. For production to be more economical, the acre yield must be increased. Already in the United States the suggestion has been made that the line of advance in crop production lies in restricting the area cultivated. A portion of the impoverished prairie lands should go back to grass. The crops needed should be raised from a smaller area. These ideas will become practi-

15

cable the moment the farmer learns how to utilize the waste products of his fields in increasing the fertility of the soil. This is the greatest need of agriculture at the present day.

BIBLIOGRAPHY

CLARKE, G.--'Some Aspects of Soil Improvement in Relation to Crop Production,' Proc. of the Seventeenth Indian Science Congress, Asiatic Society of Bengal, Calcutta,1930, p. 23.

DUCKER, H. C.--'Soil Erosion Problems of the Makwapala and Port Herold Experiment Stations, Nyasaland,' Empire Cotton Growing Review, 8, 1931, p. 1O.

FELSINGER, E. O.--'Memorandum on a System of Drainage Calculated to Control the Flow of Water on Up-country Estates, with a view to reducing Soil Erosion to a Minimum,' Tropical Agriculturist, 71, 1928, p. 221; 74, 1930, p. 68.

HOWARD, A.--Crop Production in India, a Critical Survey of its Problems, Oxford University Press, 1924.

HOWARD, A. and HOWARD, G. L. C.--The Development of Indian Agriculture, Oxford University Press, 1929.

KING, F. H.--Farmers of Forty Centuries or Permanent Agriculture in China, Korea and Japan, London, 1926.

LIPMAN, J. G.--'Soils and Men,' Proc. of the Inter. Congress on Soil Science, Washington, D.C., 1928, p. 18.

MATTHAEI, L. E.--'More Mechanization in Farming, International Labour Review, Geneva, 23, 1931, p.324.

PERCY, LORD EUSTACE--Education' at the Cross Roads, London, 1930.

Report of the Royal Commission on Agriculture in India Calcutta, 1928.

RIDDELL, W. A.--'The Influence of Machinery on Agricultural Conditions in North America,' International Labour Review, Geneva, 13, 1926, p. 309.

WAGNER, W.--Die Chinesische Landwirtschaft, Berlin, 1926, p 222.

II. ORGANIC MATTER AND SOIL FERTILITY

The ancients and the moderns are in the completest agreement as to the importance of organic matter in maintaining the fertility of the soil. This is evident when the methods of crop production in the time of the Romans are compared with the views now held by many of the leading experiment station workers in the United States and other parts of the world. In Roman times, the management of the manure heap had already reached an advanced stage. In 40 B.C. Varro drew attention to the great importance of the complete decay of manure before it was applied to the land. To bring this about, the manure heap, during the period of storage, had to be kept moist. In A.D. 90 Columella emphasized the importance of constructing the pits (in which farmyard manure was stored) in such a manner that drying out was impossible. He mentions the need of turning this material in summer to facilitate decay, and suggested that ripened manure should always be used for corn, while the fresh material could be applied with safety to grass land. The Romans therefore not only understood the importance of organic matter in crop production but had gone a long way towards mastering the principle that, to obtain the best results, it is necessary to arrange for the decay of farmyard manure before it is applied to arable land. It is interesting to turn from the writings of the ancients to the account of the symposium on 'Soil Organic Matter and Green-manuring' arranged by the American Society of Agronomy at Washington D.C. on 22 November 1928, the main results of which appeared in the Journal of the American Society of Agronomy of October 1929. Without exception, the investigators who took part in this conference laid the greatest emphasis on the importance of keeping up the supply of organic matter in the soil, and on discovering the most effective and the most economical method of doing this under the various conditions, as regards moisture, which the soils of the United States present.

During the 2,000 years which have elapsed since Varro wrote in 40 B.C. and the American investigators met in 1928, there has occurred only one brief period during which the role of organic matter was to some extent forgotten. This took place after Liebig's Chemistry in its Application to Agriculture and Physiology first appeared in 1840. Liebig emphasized the fact that plants derive their carbon from the carbon dioxide of the atmosphere and advanced the view that, in order that a soil may remain fertile, all that is necessary is to return to it, in the form of manure, the mineral constituents and the nitrogen that have been taken away in the crop. The discovery of the true origin of the carbon of plants not unnaturally suggested that the organic matter in the soil was of little consequence. Nitrogen and minerals only remained, the latter being found in the plant ashes. When therefore analyses of the crops had been made, it would be possible to draw up tables showing the farmer what he must add in the way of nitrogen and minerals in any particular case. These views and the controversies to which they gave rise, combined with the results of the Rothamsted experiments (started by Lawes and Gilbert in 1843) led to the adoption of artificial manures by many of the farmers of Europe. The Rothamsted experiments undoubtedly proved that if the proper quantities of combined nitrogen, phosphates and potash are added to the soil, satisfactory crops for many years can be obtained without the addition of organic matter beyond that afforded by the roots of the crops grown. Further, the results of hundreds of trials, in the course of ordinary farming practice, confirmed the fact that the judicious addition of nitrogenous artificial fertilizers can, in the great majority of cases, be relied on to increase the yield. It was only natural that results of this kind, combined with the important fact that the application of artificials often pays in practice, produced a marked effect on current opinion and also on teaching. For nearly a century after Liebig's ideas first appeared, the majority of agricultural chemists held that all that mattered in obtaining maximum yields was the addition of so many pounds of nitrogen, phosphorus and potassium to the acre. Beyond this the only other factor of importance was the liming of acid soils. The great development of the artificial manure industry followed as a matter of course.

THE WASTE PRODUCTS OF AGRICULTURE

The place of organic matter in the soil economy was forgotten. The old methods of maintaining soil fertility naturally fell into the background.

For a time all seemed to go well. It is only in comparatively recent years that experiment station workers have begun to understand the part played in crop production by the micro-organisms of the soil and to realize that the supply of artificials is not the whole story. Something more is needed. The need for the maintenance of the supply of organic matter soon became apparent. The view now beginning to be held is that, only after the supply of organic matter has been adequately provided for, will the full benefit of artificials be realized. There appears to be a great field for future experiment in the judicious use of artificials to land already in a fair state of fertility.

In all this however there was one important exception. In the Orient, the artificial manure phase had practically no influence on indigenous practice and passed unheeded. The Liebig tradition failed to influence the farmers of forty centuries. No demand for these products of the west exists in China. At the present day it would be difficult to purchase such a substance as sulphate of ammonia in the bazaars of rural India.

SOIL HUMUS, ITS ORIGIN AND NATURE

What is the origin and nature of the organic matter or soil 'humus' and what part does it play in soil fertility? These matters form the subject of the present chapter.

NOTE: In the presentation which follows, the fullest use has been made of (1) one of the papers of Waksman (Paper No. 276 of the Journal Series, New Jersey Agricultural Experiment Station, Department of Soil Chemistry and Bacteriology, afterwards published in Soil Science, 22, 1926, p. 123) and (2) of the symposium on soil organic matter and green-manuring which appeared in the issue of the Journal of the American Society of Agronomy of October 1929. These important contributions to the subject have made it easy briefly to sketch the necessary scientific background for the presentation of the Indore process.

THE WASTE PRODUCTS OF AGRICULTURE

The organic matter found in the soil consists of two very different classes of material: (1) the constituents of plants and animals which have been introduced into the soil and are undergoing decomposition; various unstable intermediate products which have been formed under certain environmental conditions; substances like lignified cellulose which are more resistant to decomposition and which may persist in the soil for some time; and (2) number of valuable materials which have been synthesized by the numerous groups of micro-organisms which form the soil population. The soil organic matter is thus a heterogeneous mass of substances which is constantly undergoing changes in composition. When its composition reaches a certain stage of equilibrium, it becomes more or less homogeneous and is then incorporated into the soil as 'humus'. This definition of soil organic matter, which is due to Waksman, is of great importance. Soil organic matter or 'humus' is not merely the residue left when vegetable and animal residues decay. It contains in addition the valuable materials synthesized and left behind by the fungi and bacteria of the soil population. Moreover it is a product of the general soil conditions which obtain in any particular locality, and therefore varies in composition and character from one soil type to another. It is not the same all over the world. The soil humus for example of the black cotton soils of India is not identical with that of the alluvium of the Indo-Gangetic plain.

The various steps in the formation of soil organic matter are somewhat as follows. When the fresh remains of plants or animals are added to the soil, a portion of this organic matter is at once attacked by a large number of the micro-organisms present. Rapid and intense decomposition ensues. The nature of these organisms depends on the soil conditions (mechanical and chemical composition and physical condition) and on the soil environment (moisture content, reaction and aeration, and the presence of available minerals). The decomposition processes can best be followed by measuring one of the end-products of the reaction--carbon dioxide. The rate of evolution of this gas depends on the nature of the organic matter, on the organisms which take part in the process and on the soil environmental conditions. As soon as the readily decomposable constituents of the plant and animal remains (sugars, starches, pectins, celluloses, proteins, amino-acids) have disappeared, the speed of decomposition diminishes and a condition of equilibrium tends to become established. At this stage only those constituents of the original organic matter, such as the lignins which are acted upon slowly, are left. These and the substances synthesized by the micro-organisms together form the soil humus and then undergo only a slow transformation during which a moderate but constant stream of carbon dioxide is liberated. At the same time the nitrogen of this soil humus is similarly converted into ammonia which, under favourable conditions, is then transformed into nitrate. It will be clear therefore that the soil organic matter or humus is a manufactured product and that its composition is not everywhere the same, but will vary with the soil conditions under which it is produced. Like all manufactured articles, it must be properly made if it is to be really effective. Too much attention therefore cannot be paid to its preparation.

After the production of humus and its incorporation into the soil mass, the next step is its utilization by the crop. This can only take place when this organic matter is decomposed by the micro-organisms of the soil. This process is very slow, as can be seen by placing a quantity of soil under favourable environmental conditions and measuring the rate of decomposition, either by the evolution of carbon dioxide or by the accumulation of ammonia and nitrate nitrogen. Since the ratio between the carbon and nitrogen content of the humus in normal cultivated soils is more or less constant, approaching 10:1, the evolution of carbon dioxide will be accompanied by the liberation of available nitrogen. This oxidation of the carbon and of the nitrogen is comparatively very slow, as only slow-growing groups of microorganisms are capable of attacking it. These organisms are aerobic and moreover can only work effectively when the general soil reaction is favourable. Their activities are therefore hastened in non-acid peat soils by draining, in acid peat soils by draining and liming, and in acid soils by liming.

It will be clear that the utilization of vegetable and animal wastes in crop production involves two definite steps: (1) the formation of humus and its incorporation into the soil and (2) the slow oxidation of this product accompanied by the production of available nitrogen. Both of these stages are brought about by micro-organisms for which suitable environmental conditions are essential. The requirements of the first phase--the preparation of humus and its incorporation into the soil mass--are so intense that if the process takes place in the soil itself, it is certain to interfere with the development of the crop. The needs of the second phase--the utilization of humus--are much less intense and can proceed in the soil without harm to the growing plant. From the point of view of crop production therefore, it will be a distinct advantage to separate these two stages and to prepare the humus outside the field. In this matter the Chinese have anticipated the teachings of western science. The cultivators of the Orient were the first to grasp and act upon the master idea that the growth of a crop involves two separate processes, the preparation of food-materials from vegetable and animal wastes which must be done outside the field, and the actual growing of the crop. Only in this way can the soil be protected from overwork

THE FORMATION OF HUMUS AS A RESULT OF THE SYNTHESIZING ACTIVITIES OF MICRO-ORGANISMS

Although the important part played by microorganisms in the formation of soil humus has only very recently been fully understood, nevertheless the older literature contains a number of useful contributions to the subject. Most of these early papers appeared towards the end of the last century; many of them related to other branches of knowledge and were not written from the point of view of agriculture. They have been summed up by Waksman, from whose paper the following account has been prepared. Post-Ramann and Muller considered that the 'humus' bodies obtained from soil often consist of the chitinous remains of insects and animal excrete. Wettstein and Winterstein showed that chitin is characteristic of various fungi and not of bacteria. Schmook advanced the view that the protein nitrogen in the soil was mostly present in the bodies of bacteria and protozoa. Trussov showed that the protoplasm of fungi is a source of humus in the soil. Schreiner and Storey suggested that various characteristic constituents of the soil are probably synthesized by micro-organisms.

The earlier work on this subject has been considerably developed, first by Falck and more recently by Waksman. Falck showed that organic matter in forest soils can be transformed into different types of humus in at least three ways: (1) The yearly additions of raw organic matter are completely decomposed by fungi (microcriny) accompanied by the synthesis of fungus protoplasm, which serves as an excellent fertilizer for the forest trees. In this process the celluloses are decomposed completely, whereas the lignins are more resistant. (2) The decomposition of the organic matter is begun by fungi and then carried on by lower invertebrates and bacteria (anthracriny). The fungus mycelium as well as the original organic matter are devoured by various larvae producing a dark 'humus' mass which, in the presence of bases, is oxidized by bacteria with the ultimate liberation of carbon dioxide and the formation of nitrate. (3) The formation of peat (anthrogeny), which Falck explains as resulting from the absence of an abundant fungus development. Waksman carried the subject still further and called attention to the similarity between the carbon-nitrogen ratio of the soil organic matter and that of the protoplasm of the soil fungi and other micro-organisms, and suggested that these probably make up a large part of the soil 'humus'. He further pointed out that when cellulose is added to the soil, it decomposes only in proportion to the available combined nitrogen present. This is because the decomposition is brought about by fungi and bacteria, both of which require combined nitrogen. The ratio between the amount of cellulose decomposed and the nitrogen required is about 30:1, so that, for every thirty parts of cellulose decomposed by the fungi and bacteria, one part of inorganic nitrogen (ammonium salt or nitrate) will be built up into microbial protoplasm. In the presence of sufficient combined nitrogen and under aerobic conditions, the decomposition of cellulose is very rapid. The same is true of vegetable wastes like straw, maize stalks, wood products and other materials rich in celluloses, pentosans and lower carbohydrates but poor in nitrogen. These facts explain the injurious effects on crop growth which follow the addition of straw and green-manure to the soil. The decomposition of these materials removes large quantities of combined nitrogen from the soil solution. This nitrogen is then temporarily stored in the form of microbial protoplasm, when for a time it is placed beyond the reach of the growing crop.

Since Waksman's paper appeared in 1926, an important contribution to this subject has recently been made by Phillips, Weite and Smith. The results of these investigators (which agree with our experience at Indore) has removed the impression that lignin is comparatively resistant to the action of micro-organisms. Under suitable conditions, soil organisms are capable of decomposing lignin as found in lignified plant materials (cornstalks, oat hulls, corn cobs and wheat straw), the rate of decomposition being as great as that of cellulose and pentosans.

THE ROLE OF HUMUS IN THE SOIL

24

From the immediately practical point of view, the actual role of humus in the soil is of even greater interest than its formation, nature and decomposition. This material influences soil fertility in the following ways:--

1. The physical properties of humus exert a favourable influence on the tilth, moisture-retaining capacity and temperature of the soil as well as on the nature of the soil solution.

2. The chemical properties of humus enable it to combine with the soil bases, and to interact with various salts. It thereby influences the general soil reaction, either acting directly as a weak organic acid or by combining with bases liberating the more highly dissociating organic acids.

3. The biological properties of humus offer not only a habitat but also a source of energy, nitrogen and minerals for various micro-organisms.

These properties--physical, chemical and biological--confer upon humus a place apart in the general work of the soil including crop production. It is not too much to say that this material provides the very basis of successful soil management and of agricultural practice.

THE WASHINGTON SYMPOSIUM ON SOIL ORGANIC MATTER

Once the origin and nature of the soil organic matter is understood and the importance of this material in soil fertility is appreciated, the next step is to consider how best to make use of this information and to weld it into farming practice. With this object in view a symposium on soil organic matter and green-manuring was arranged at Washington D.C. on 22 November 1928, when the following papers were read and discussed:--

I. 'The Relation of Soil Type to Organic Matter.' C. F. Marbut.

2. 'Organic Matter Problems in Humid Soils.' T. Lyttleton Lyon.

3. 'Organic Matter Problems Under Dry-Farming Conditions.' J. C. Russell

4. 'Organic Matter Problems in Irrigated Soils.' P. S. Burgess.

5. 'Chemical and Microbiological Principles Underlying the Use of Green-Manures.' S. A. Waksman (by title only).

6. 'Influence of Organic Manures on the Chemical and Biological Properties of Arid Soils.' J. E. Greaves.

7. 'Green-Manuring and Its Application to Agricultural.' A. J. Pieters and Roland McKee.

In dealing with the question of organic matter in humid soils, Lyon first presented a critical survey of the literature dealing with the losses of nitrogen in soils and concluded that:--

1. The loss of gaseous nitrogen may, under some conditions, cause a greater removal of nitrogen from a soil than occurs through absorption by crop plants.

2. The conditions which favour a large loss of this kind are: (a) tillage or stirring the soil in any way, (b) absence of plant growth, (c) high nitrogen content of a soil, (d) application of large quantities of nitrogenous manures, and (e) possibly the application of lime to some soils.

3. The loss of gaseous nitrogen does not take into account the amount fixed by soil organisms and therefore the calculated losses are less than actually occurred.

These losses of gaseous nitrogen from the soil may arise in five possible ways:--

1. There may be an escape of part of the ammonia during the process of ammonification.

2. There may be a reduction of nitrates to form nitrogen as a result of alternating oxidation and reduction.

26

3. There may be a loss of gaseous nitrogen in the oxidation of ammonia to nitrous acid since nitrogen is possibly an intermediate product in this process.

4. A loss of nitrogen may result from the interaction of nitrous acid with the NH2 group of the amino-acids.

5. A loss of gaseous nitrogen may occur as a result of the decomposition of ammonium nitrite in the process of nitrification.

In connexion with these losses of nitrogen it was pointed out in the discussion that the following two facts must be considered: (1) The ratio of carbon to nitrogen in the soils of the humid regions tends to maintain itself in the region of 10:1. If the organic residues left in the soil or applied to it afterwards have a higher carbon-nitrogen ratio than 10:1, an adjustment is soon effected, the extra carbon disappearing into the atmosphere as carbon dioxide. If the carbon-nitrogen ratio is less than 10:1, there is likely to be a loss of nitrogen before the ratio is adjusted. (2) The nitrogen content of any given soil tends to come to an equilibrium at a point which depends upon the nature of the soil, the effective climate and the cropping system. When therefore the nitrogen supply is increased in any way, the excess is soon dissipated when the soil comes under cultivation.

The information placed before the meeting by Russel (Nebraska) on the role of organic matter under dry-farming conditions was most instructive, and throws a flood of light on the consequences which are certain to follow the continuous cropping of virgin land without manure. A rapid and continuous fall in the total organic matter content, accompanied by loss of nitrogen, occurs together with a corresponding falling off in cropping power. Side by side, the water-holding capacity of these soils decreases, while the structure and tilth exhibit marked degeneration. All this has naturally led to attempts being made to restore the original content of organic matter. The results obtained, however, have been most disappointing, for the reason that most of these efforts have been directed towards the direct incorporation of green-manures and raw organic matter like straw into the soil under conditions of low rainfall. In many cases more harm than good has resulted. Russel concludes that the problem of the restoration of organic matter under dry-land conditions is extremely complicated and

27

difficult and leans to the view that the solution of the problem might after all be found in the direction of nitrogenous fertilizers. Experience at Indore, however, suggests that all these difficulties could at once be avoided if the available supplies of green-manure, straw and other raw organic matter could first be composted outside the field before being applied to the land. The American farmers are obviously trying to overwork the soil and Mother Earth naturally objects.

The application of organic matter to the soil is followed by a number of important indirect results. These were dealt with by Greaves in a most interesting communication, in which the results obtained over a number of years on two different types of Utah soils were discussed. The first (Nephi) was typical dry-farm soil, the second was under irrigation (Greenville). In both the results were similar. The application of organic matter increased the ammonifying, nitrifying and nitrogen fixing processes of the soil. The gains in nitrogen, due to non-symbiotic nitrogen fixers, occurring under greenhouse conditions, varied from 0 to 304 lb. per acre foot of soil. The greatest gains occurred when legumes were used in the manure. The gain occurring in the soil under field conditions, and attributed to non-symbiotic nitrogen fixation, was 44 lb. per acre annually. Approximately 3,000 lb. of applied organic material were decomposed every year.

The last paper of the symposium dealt with the practice of green-manuring throughout the United States, with the various crops which are turned under, and with the great need for further exact experimentation on this question. Pieters and McKee state: 'In reviewing the experimental work that has been done with green-manures in the United States and the practices that are now followed it is evident that much work remains to be done before many questions can be settled or answered. Some of these fall clearly in the field of chemistry, others in physiology, and still others in bacteriology or other specialized fields of biology. Some, however, are strictly agronomic problems or so directly involved with crop production that their solution can perhaps best be undertaken by the agronomist or carried on with his active cooperation. It takes but a hasty survey to indicate the wide scope this work must cover in order to answer the specific questions for the many soil types, various climatic conditions, and for each of the large number of agronomic and horticultural crops involved.' In no case is there any reference in this paper to the growing of green-manures for the express purpose of providing material for composting, possibly because the need for this material has not yet been fully realized and because of the labour involved. Green-manuring in the United States, as in India and other parts of the world, is still in the empirical stage. Green crops are grown merely to provide a supply of organic matter for turning into the soil. What happens afterwards is a matter of chance. If the results are favourable, so much the better; if anything untoward occurs, one must hope for better things next time. That such an uncertain practice persists at all in the United States and that it appears to be spreading can only be explained by the great need of these depleted soils for fresh supplies of organic matter.

BIBLIOGRAPHY

LIEBIG, J.--Chemistry in its Application to Agriculture and Physiology, 1840.

PHILLIPS, M., WETTE, H. D., and SMITH, N. R.--The Decomposition of Lignified Materials by Soil Microorganisms,' Soil Science, 30, 1930, p. 383.

RUSSELL, E. J.--Soil Conditions and Plant Growth, London, 1927.

RUSSELL, E. J. and RICHARDS, E. H.--'The Changes taking place during the Storage of Farmyard Manure,' Journ. Of Agric. Science, 8, 1917, p. 495.

'Symposium on Soil Organic Matter and Green-Manuring,' Journ. of the American Society of Agronomy, 21, 1929, p. 943

WAKSMAN, S. A.--'The Origin and Nature of the Soil Organic Matter or Soil "Humus": 1--Introductory and Historical,' Soil Science, 22, 1926, p. 123.

III. THE SOURCES OF ORGANIC MATTER

A number of sources of soil organic matter exist, namely: (1) the roots of crops left behind at harvest, including the weeds turned under in the course of cultivation; (2) the algae met with in large quantities in rice fields, on the surface of the soils of tropical countries during the rainy season and to some extent in all soils; (3) green-manure; (4) farmyard manure; (5) artificial farmyard manure. In addition to these supplies, certain by-products of industries, such as oil-cakes and wool-waste, are also employed as sources of organic matter. These, however, are small in total amount and need not be considered. Except in China and Japan and to a limited extent in India, little or no use is made of night soil in crop production.

THE ROOT-SYSTEMS OF CROPS

It is not always realized that about half of every crop--the root-system--remains in the ground at harvest time and thus provides automatically a continuous return of organic matter to the soil. The weeds and their roots turned in during the ordinary course of cultivation add to this supply. When these residues, supplemented by the fixation of nitrogen from the atmosphere, are accompanied by skilful soil management, crop production can be maintained at a moderate level without the addition of any manure whatsoever. A good example of such a system of farming without manure is to be found on the alluvial soils of the United Provinces, where the field records of ten centuries prove that the land produces fair crops year after year without any falling off in fertility. A perfect balance has been reached between the manurial requirements of the crops harvested and the natural processes which recuperate fertility. A similar, although not so striking a result, is afforded by the permanent wheat plot at Rothamsted, where this crop has been grown every year on the same land without manure since 1844. This plot, which has been without manure of any kind since 1839, showed a slow decline in production for the first eighteen years after which the yield has been practically constant. Systems of soil management such as these provide, as it were, the base line for the would-be improver. Nothing exists in the world's agriculture below this level. At the worst, therefore, the organic matter of a soil, constantly cropped without manure, does not disappear altogether. The wheel of life slows down. It does not stop.

SOIL ALGAE

One source of readily decomposable organic matter, which is available in India just at the moment when the cold season crops need it, is to be found in the shape of a thick algal film on the surface of cultivated soils during the second half of the rains. This film has also been observed in Africa, Ceylon and Java, and is probably universal during the rainy season in all parts of the tropics. As is well known, there are two periods in India when the crop is in greatest need of combined nitrogen: (1) at the break of the monsoon in June and July, and (2) when the cold season crops are sown in October after the rains. These latter are planted at a time when the available nitrogen in the surface soil is likely to be in great defect. The land has been exposed to heavy rain for long periods; the surface soil is often waterlogged. Nitrates under such conditions are easily lost by leaching and also by de-nitrification. The conditions are therefore altogether unfavourable for any approach towards an ample supply of nitrate when sowing time comes round in early October. How do the cold weather crops obtain a sufficient supply of this essential food material? It is more than probable that the deficiency is made up for, in part at least, by the rapid decay of the algal film (which also appears to be one of the factors in nitrogen fixation) during the last cultivations preceding the sowing of the cold weather crop in October. It is possible that some changes may have to be made in soil management with a view to stimulating the growth of this algal film. One of the beneficial effects of growing a green-manure crop like sann hemp for composting, during the early rains, may prove to be due to the favourable environment provided for the rapid establishment of the algal film. On monsoon fallow land it will probably be found best to suspend surface cultivation during the

31

second half of the rains when the film is most active. There is already among the cultivators of India a tendency to stop stirring the surface, from the middle to the end of the rains, even when this involves the growth of weeds. This coincides with the period when the algal film is most noticeable. The indigenous practices may therefore prove to be based on sound scientific principles. Here are ready to hand several interesting subjects which urgently call for study under actual tropical conditions. When this is undertaken, the investigation should include: (1) the conditions most favourable for the establishment of the algal film; (2) the part played by algae and associated bacteria in nitrogen fixation; (3) the role of algae in banking easily destroyed combined nitrogen during the rains; and (4) the supply of easily decomposable and easily nitrifiable organic matter for the use of the cold weather crops. In the rice fields of the tropics, the algal carpet is even more evident than on ordinary cultivated soils. The total weight of organic matter added every year to each acre of rice land in the shape of algal remains must be considerable and must serve as a useful addition to the store of organic matter. Apart from the fixation of nitrogen from the air, it may help to explain why such heavy crops of paddy can be obtained in India, year after year on the same land, without manure.

GREEN-MANURES

Since the investigations of Schulz-Lupitz first showed how open sandy soils in Germany can be rapidly improved in texture by the incorporation of green-manures, the future possibilities of this method of enriching the land became apparent to the investigators of the Occident. After the role of the nodules (found on the roots of leguminous plants) in the fixation of atmospheric nitrogen was proved, the problems of green-manuring have naturally centred round the utilization of the leguminous crop in adding to the store of organic matter and combined nitrogen in the soil. At the end of the last century it seemed so easy, by merely turning in a leguminous crop, to settle at one stroke and in a very economical fashion the great problem of maintaining soil fertility. At the expenditure of a very little trouble, the soil might be made to manure itself. A supply of combined nitrogen, as well as a fair quantity of organic matter, might be provided without any serious interference with ordinary cropping. These expectations have led to innumerable green-manuring experiments all over the world with practically every species of leguminous crop. The results however have left much to be desired. In a few cases, particularly on open soils and where the rainfall, after the ploughing in of the green crop, is well distributed, the results have been satisfactory. On rice lands, where abunabundance of water ensures the maintenance of swamp conditions, somewhat similar results have been obtained. In the vast majority of cases, however. green-manuring has been disappointing. As a general method of soil improvement, the game is hardly worth the candle. On the monsoon fed areas of India the rainfall is often so uncertain, after the green crop is ploughed in, that for long periods decay is arrested. Sowing time arrives at a stage when the soil contains a mass of half-rotted material, with insufficient nitrogen and moisture for the growth of a crop. Failure results. The crops raised after green-manure are worse than those obtained on similar land left fallow. For this reason green-manuring has not been taken up by the people in India, in spite of the experiments and propaganda of the Agricultural Department.

It soon became evident, during the early years of the present century in India, that no matter what the rainfall and the soil conditions may be, a definite time factor is in operation in green-manuring. A period of not less than eight weeks must elapse, between the ploughing in of the green crop and the planting of the next, if satisfactory results are to be obtained. This was well brought out in the green-manuring experiments on tobacco, carried out at Pusa between 1912 and 1915. Some years later, the explanation of this factor, as well as the general conditions necessary for the decay of a green-manure crop were furnished by the work done at the New Jersey experiment station by Waksman and his co-workers. The decay and incorporation of green-manure in the soil has been shown to be a very complex process, depending on: (1) the chemical composition of the plants which make up the green-manure, which in turn largely depends on the age of the crop when ploughed in; (2) the nature of the decomposition of the various groups of organic complexes in the plant by the different types of soil organisms, which in turn is influenced by such factors as moisture, aeration, and the supply of available nitrogen and phosphates needed by these organisms, and (3) the metabolism of the microorganisms taking part in the decay of the green crop.

The process of incorporation takes place on the following lines. When the green-manure crop is ploughed in, the first stages of decay are brought about by fungi, which require for their activities ample supplies of air, moisture and combined nitrogen, as well as the soluble and easily decomposable carbohydrates supplied by the green crop. If the supply of nitrogen provided by the green-manure is insufficient, the stores of soluble nitrates in the soil solution are utilized by the fungi. Decay is rapid provided all these essential factors are simultaneously arranged for. The result is that the whole energies of the soil at this period are given up to the needs of the fungi of decay, which synthesize large quantities of protoplasm from the materials supplied by the green crop and the soil solution. During this phase, most of the nitrogen present is built up into mycelial tissue, and is therefore not immediately available for the growth of crops. The next stage is the decay of the remainder of the green-manure, including the mycelial tissue itself, by various groups of bacteria, followed by the incorporation of the whole mass into the soil organic matter. This must first be nitrified before the soil solution and the crop can obtain any benefit from this form of manuring. Clearly all this takes time, and needs abundance of oxygen as well as a continuous supply of soil moisture. If any of the limiting factors--nitrogen supply, air or moisture--are in defect, it is obvious that the final stage of nitrifiable organic matter will not be quickly reached. The soil will not only contain a mass of undigested material, but will be poor in available nitrogen and perhaps low in moisture as well. Seeds sown in such a soil can only result in a poor crop. The investigations of the New Jersey experiment station explain the importance of the time-factor in green-manuring, and incidentally show that the ordinary green-manuring experiments in India cannot possibly succeed. The sooner they are discontinued the better. Nothing is to be gained by attempting the hopeless task of manufacturing soil organic matter under conditions which cannot be controlled.

The question at once arises as to whether the green-manuring process can be regulated in such a manner that the results can be relied upon? A number of attempts have been made in this direction in India, of which that carried out by Clarke at Shahjahanpur is the most promising. Green crops of sann hemp (Crotalaria juncea L.) have been successfully utilized for the growth of sugar-cane. The secret of the Shahjahanpur process is to provide ample moisture, by means of irrigation, for the first stages of the decay of the green-manure. The rainfall, after the hemp crop is ploughed in, is carefully watched. If it is less than five inches during the first fortnight of September, the fields are irrigated. This enables the first phase of the decay of the green crop by the soil fungi to be completed. Practically all the nitrogen is then in the form of easily decomposable mycelial tissue. During the autumn, nitrification is prevented by drying out the surface soil. The nitrogen is, as it were, kept in the bank till the sugar-cane is planted under irrigation in March. Nitrification then sets in and the available supplies of combined nitrogen are made use of by the sugar-cane. In this way crops of over thirty tons of cane to the acre have been grown without the addition of any manure beyond the hemp, grown on the same land the previous rains and treated in the manner indicated above. These results do not appear to have been obtained with any other crop than sugar-cane planted in March. It would be interesting to have figures for wheat, sown in October, i.e. about six weeks after the hemp was ploughed under. It is probable that even with irrigation, this interval is insufficient for the proper incorporation of the green crop into the body of the soil organic matter and its subsequent nitrification. In this case, the Shahjahanpur method, valuable and interesting as it is, can only have a limited application.

Is it possible to devise a method of green-manuring, by means of the leguminous crop, which avoids all risks, is certain, and also makes the fullest use of this system? There are two possible ways in which the growing of a leguminous green-manure crop may benefit the soil. These are: (1) the well-known advantages of such crops in the rotation in increasing the nitrogen supply and in stimulating the micro-organisms in the soil, and (2) the effects of incorporating the green crop into the store of soil organic matter. Lohnis, however, showed, in many green-manuring experiments with leguminous crops, that the same results were obtained when the crop was removed as when it was ploughed under--a conclusion which is in full accord with Waksman's work. It follows from this that the double advantage of a leguminous green-manure crop can only be achieved provided fall use of the crop itself can be found outside the field, either as fodder for animals, for making silage or as material for the manufacture of compost. This latter method has been successfully worked out at Indore, and will be described in the next chapter. The real place of the leguminous crop in green-manuring seems to be in providing material for the manufacture of organic matter in a compost factory, specially designed for the purpose.

The exact period in the life history of the green crop, when it should be reaped for composting, is an important matter. If the crop is cut before the grand period of growth is completed, the maximum amount of vegetable waste will not be obtained. On the other hand, an early harvest will yield a product rich in nitrogen and suitable for rapid decay (Appendix C). Late harvesting is also attended with disadvantages. If reaped after flowering begins, the green crop will have used up a good deal of the rich nodule tissue which will then be temporarily removed from the soil and will not benefit the next crop. Further, the older the crop, the more unfavourable the carbon-nitrogen ratio becomes. The best stage for removal will be just before flowering begins. At this point, most of the nitrates in the soil solution have been absorbed by the crop and have been banked, either in the form of an easily decomposable root-system or as compost material, the chemical composition of which is exactly what is needed to improve the carbon-nitrogen ratio of the other vegetable wastes of the farm. When the green crop is reaped at this stage the following advantages are obtained: (1) The nitrates of the soil solution are safely banked. (2) The next crop derives the maximum benefit from an easily decomposable and uniformly distributed root-system, rich in combined nitrogen, the decay and incorporation of which is well within the powers of the soil. (3) The store of vegetable waste for composting is increased in amount and improved in chemical composition by the uniform distribution of the combined nitrogen throughout the tissues of the green crop.

FARMYARD MANURE

From the beginning of agriculture, the utilization of farm wastes, rotted by means of the urine and dung of animals, has been the principal means of replenishing soil losses. Even at the present day, in spite of the establishment of numerous experiment stations and the employment of an army of investigators, the methods in vogue in the preparation and storage of this product leave much to be desired. Even under the covered-yard system, when the dung and litter are left under the animals until a layer several feet thick is produced, and the product is protected from the weather, as much as fifteen per cent of the valuable nitrogen is lost. When the dung is carted out into a heap to ripen, as is the usual practice, the losses of nitrogen are even greater. Russell and Richards, who some years ago carried out an elaborate investigation on the storage of farmyard manure at Rothamsted, concluded that: (1) the system of leaving the manure under the beasts till it is required for the fields, as in the box or covered-yard system, is the best whenever this is practicable; (2) the ideal method of storage is under anaerobic conditions at a temperature of 26 degrees C.; (3) the manure heap, however well made and protected, involves losses of nitrogen; and (4) the best hope of improvement lies in storing the manure in watertight tanks or pits, so made that they can be completely closed and thereby allow the attainment of perfect anaerobic conditions. These investigations, published in 1917, clearly indicate that one of the reasons for the present imperfect management of farmyard manure lies in the fact that the conditions are sometimes aerobic, at others anaerobic, whereas they should be one or the other throughout. In other words, there is no proper management of the air supply. Moisture is not usually in defect, except in hot countries like India where there is abundant air but often little moisture. Taking Great Britain and India as extreme cases of the management of farmyard manure, we find one or other of the following conditions in operation. In Great Britain, the irregular air supply of the manure heap leads to serious losses of nitrogen.

The final product is not a fine powder but a partially rotted material, which cannot be incorporated into the pore-spaces of the soil until further decay has taken place. The soil therefore has to do a good deal of work before the farmyard manure, applied on the surface in lumps, can be uniformly distributed through and incorporated into the soil mass. In India, the storage of farmyard manure leads to the loss of so much moisture, that often insufficient decay takes place before it finds its way into the soil. Losses of nitrogen may be prevented in this way but the work thrown upon the soil is even greater than in temperate regions. Only in China and Japan is any real attempt made to prepare the manure for the use of the crop, and to relieve the soil from unnecessary work. What is needed throughout the world is a continuous system of preparing farmyard manure in which (1) all losses of nitrogen are avoided, and, (2) the various steps from the raw material to the finished product follow a definite plan, based on the orderly breaking down of the materials, and the preparation of a finished product, ready for immediate nitrification, which can easily be incorporated into the soil. At the same time, an attempt should be made to gain as much nitrogen as possible by fixation from the atmosphere. Only when all this is done will the preparation of farmyard manure be based on correct scientific principles.

ARTIFICIAL FARMYARD MANURE

During the last ten years, an additional source of soil organic matter has been utilized, namely, artificial or synthetic farmyard manure. In 1921, the results of experiments, carried out by Hutchinson and Richards at Rothamsted on the conversion of straw into manure without the intervention of live stock, were published. In this pioneering work, which constitutes an important milestone in the development of crop production, a method was devised by which straw could be converted into a substance having many of the properties of stable manure. In the preliminary experiments, the most promising results were obtained when the straw was subjected to the action of a culture of an aerobic cellulose decomposing organism (Spirochoeta cytophaga), whose activities were found to depend on the mineral substances present in the culture fluid. The essential factors in the production of well-rotted farmyard manure from straw were found to be: air supply; a suitable temperature, and a small amount of soluble combined nitrogen. The fermentation was aerobic; the breakdown of the straw was most rapid in a neutral or slightly alkaline medium in the presence of sufficient available nitrogen. Urine, urea, ammonium carbonate and peptone (within certain concentrations) were all useful forms of combined nitrogen. Sulphate of ammonia by itself was not suitable, as the medium soon became markedly acid. The concentration of the combined nitrogen added was found to be important. When this was in excess, nitrogen was lost from the mass before decay could proceed; when it was in defect, a marked tendency to fix nitrogen was observed. The publication of this paper soon led to a number of further investigations, and to numberless attempts all over the world to prepare artificial farmyard manure from every kind of vegetable waste. The principles underlying the conversion are now well understood, and have recently been summed up by Waksman and his co-workers in the Journal of the American Society of Agronomy (21, 1929, p. 533) in a paper which should be carefully studied by all interested in this important subject. The principles underlying the conversion are so well put by these investigators that they are best given in the authors' own words:--

'The problems involved in the study of the principles underlying the decomposition of mature straw and other plant residues in composts, leading to the formation of so-called artificial manure, involve a knowledge of: (a) the composition of the plant material; (b) the mechanism of the decomposition processes which are brought about by the micro-organisms; and (c) a knowledge of the metabolism of these organisms.

'Straw and other farm residues, which are commonly used for the purpose of composting, consist predominantly (60 per cent or more) of celluloses and hemi-celluloses, which undergo rapid decomposition in the presence of sufficient nitrogen and other minerals, of lignins (15 to 20 per cent) which are more resistant to decomposition and which gradually accumulate, of water-soluble substances (5 to 12 per cent) which decompose very rapidly, of proteins which are usually present in very small amounts (2.2 to 30 per cent) but which gradually increase in concentration with the advance of decomposition, and of the mineral portion or ash.

'The processes of decomposition involved in the composting consist largely in the disappearance of the celluloses and hemi-celluloses, which make up more than 80 per cent of the organic matter which is undergoing decomposition in the process of formation of artificial manures. These poly-saccharides cannot be used as direct sources of energy by nitrogen-fixing bacteria and their decomposition depends entirely upon the action of various fungi and aerobic bacteria. In the process of decomposition of the celluloses and hemi-celluloses, the micro-organisms bring about the synthesis of microbial cell substance. This may be quite considerable, frequently equivalent to a fifth or even more of the actual organic matter decomposed. To synthesize these large quantities of organic matter, the micro-organisms require large quantities of available nitrogen and phosphorus and a favourable reaction. The nitrogen and phosphorus are used for the building up of the proteins and nucleins in the microbial cells. Since there is a direct relation between the celluloses decomposed and the organic matter synthesized, it should be expected also that there would be a direct relation between the cellulose decomposed and the amount of nitrogen required. As a matter of fact, for every forty or fifty parts of cellulose and hemi-cellulose decomposed, one unit of available nitrogen has to be added to the compost.

'As the plant residues used in the preparation of "artificial manure" are poor in nitrogen, available inorganic nitrogen must be introduced for the purpose of bringing about active decomposition.

This explains the increase in the protein content of the compost accompanying the gradual decrease of the celluloses and hemicelluloses.

'In general, artificial composts can be prepared from plant residues of any chemical composition so long as the nature of these residues and of the processes involved in their decomposition are known. By regulating the temperature and moisture content and by introducing the required amounts of nitrogen, phosphorus, potassium and calcium carbonate, the speed of decomposition and the nature of the product formed can be controlled.'

It is not possible in the space available to summarize all the various experiments which have been made in Great Britain, the United States, India and other parts of the world on the actual conversion of vegetable residues into artificial farmyard manure. It will be sufficient to refer to typical examples of what has been done. The Rothamsted investigations have been continued and have led to a patented process, known as Adco, by which the requisite nitrogenous and phosphatic food for the micro-organisms, as well as a base for the neutralization of acidity, are added to the vegetable wastes in the form of powders. Full details and numerous illustrations are to be found in the various Adco pamphlets. The object of patenting the process is not profit for the inventors but the raising of funds for further research. All users of Adco therefore are not only provided with a useful mixture but also make a small contribution to the cost of fundamental research work. In India, the various experiments on the production of artificial farmyard manure from a large number of materials, such as prickly pear, fallen leaves, town refuse, mahua (Bassia latifolia L.) flowers, weeds, banana waste, leguminous plants such as sann hemp, green pea stalks and various weeds have recently been summed up by Fowler, whose paper (see Bibliography below) should be consulted for details. The materials employed for adding the necessary nitrogen and other materials for the micro-organisms were night-soil, cow-dung, cattle urine, activated sludge or chemicals like sulphate of ammonia and calcium cyanamide. A large number of experiments are described from which it is clear that very useful manures, containing from 1 to 4 per cent of nitrogen, were obtained, which in field trials with rice and maize gave results equal to or better than any other nitrogenous manure in common use. Attempts were made in the course of this work to determine the amount of nitrogen fixation from the air which occurs during the conversion of the vegetable waste. It was found, when proper care was taken to supply the necessary organisms, that a considerable amount of free nitrogen was actually absorbed. These results, which agree with others on the same point, are of considerable interest. If in the conversion of vegetable wastes into artificial farmyard manure additional nitrogen can be gained, obviously the ideal conditions have been discovered. Once such principles have been correctly ascertained and put into practice, it might then be possible to deal not only with the manure heap itself but also with green-manuring, so that actual fixation can be substituted for the losses of nitrogen which now occur.

As is to be expected in such a matter as this, the preparation of artificial farmyard manure has been in actual operation centuries before Hutchinson and Richards began their work at Rothamsted. King, in Farmers of Forty Centuries, describes the conversion by the Chinese peasants of clover (Astragalus sinicus) into manure by mixing the green crop with rich canal mud To all intents and purposes, this system closely resembles the Adco process. Once more the empirical methods, discovered during centuries of practice, have preceded the results obtained by the application of pure science. Nevertheless, although in a sense the Rothamsted workers have been anticipated, it is quite safe to say that but for their work, the utilization of green clover in China, although described in the literature of the subject, would have passed unheeded. It was the novelty of the Rothamsted investigations which has proved so useful and so stimulating.

A critical examination of the literature on the principles underlying the conversion into humus of the chief groups of crude organic matter--green-manure, farmyard manure and vegetable wastes--reveals one fundamental weakness, namely, the fragmentation, into a number of loosely related sections, of what is essentially one subject. Farmyard manure, green-manure and the preparation of synthetic farmyard manure are always dealt with as if they were separate things and not parts of one great project. Even Waksman (whose contributions to the principles underlying the conversion of vegetable wastes into humus cannot fail to compel the admiration of all investigators), when the time came to write up his work for the agronomists of the United States, contributed three separate papers to the Journal of the American Society of Agronomy--one on farmyard manure, one on green-manure and the third on artificial farmyard manure--instead of synthesizing all these related subjects into one single contribution. When we come to the practical side of the question, a similar fragmentation is apparent. Green-manuring is always a separate process. The manure heap and its utilization from the time of the Romans to the present day, forms a special section of the work of the farm. The manufacture of artificial farmyard manure is again split off as an isolated operation. This particularism, in the most recent papers, is reflected in the separate conversion of each kind of vegetable waste, although it follows, from considerations of chemical composition, that a mixture of residues is much more likely to possess a suitable carbon-nitrogen ratio than any single material. As will be evident from a study of Waksman's three papers referred to above, the principles underlying the decay of farmyard manure, of green-manure and the preparation of artificial farmyard manure are essentially the same, namely, the synthesis of humus, by means of fungi and bacteria, from crude vegetable matter, various nutrients, air, water and bases. This humus increases the supply of soil organic matter and is capable of rapid nitrification. What is needed is the welding of all the separate fragments of the subject into a well ordered system. One process is required, not several. The agriculturist of the future must be shown how to become a chemical manufacturer. Further, the method finally adopted must be so elastic that it can be introduced into almost any system of agriculture. Again, it must be simple, safe and must yield a continuous and uniform product, capable of being instantly utilized by the crop. No waste of valuable nitrogen should occur at any stage. If possible, matters should be so arranged that the fixation of atmospheric nitrogen takes place at all stages of the process--in the compost factory and afterwards in the soil. In the next chapter, a continuous process of making humus is described which fulfils the conditions just outlined. This includes, in a single process, the various fragments of the subject, such as the care of the manure heap, green-manuring, the utilization of all vegetable wastes as well as the urine earth from the cattle shed and the wood ashes from the labourers' quarters. By its means, the waste products of 300 acres of land are converted every year into about 1,000 cart-loads of valuable humus, of uniform chemical composition and of uniform fineness. When this material is added to the soil there is a rapid increase in fertility. The practical results obtained at Indore prove that all that is needed to raise crop production to a much higher level throughout the world is the orderly utilization of the waste products of agriculture itself.

BIBLIOGRAPHY

BRISTOL, B. M.--'On the Alga-flora of some desiccated English Soils: an Important Factor in Soil Biology,' Annals of Botany, 34, 192O, P.35.

BRISTOL, M. B. and PAGE, H. J.--'A Critical Enquiry into the Alleged Fixation of Nitrogen by Green Alga,' Annals of Applied Biology, 1O, 1923, p. 378.

BRISTor-ROACH, B. M.--'The Present Position of our Knowledge of the Distribution and Functions of Alga, in the Soil,' Proc. of the Inter. Congress of Soil Science, Washington, D.C., 1928, p. 30.

CARBERY, M. and FINLOW, R. S.--'Artificial Farmyard Manure,' Agric. Jonrn. of India, 23, 1928, p. 80.

CLARKE, G., BANERJEE, S. C., NAIB HUSAIN, M., and QAYUM, A.--'Nitrate Fluctuation in the Gangetic Alluvium and Some Aspects of the Nitrogen Problem in India,' Agric. Journ. of India, 17, 1922, p. 463.

CLARKE, G.--'Some Aspects of Soil Improvement in relation to Crop Production,' Proc. of the Seventeenth Indian Science Congress, Asiatic Society of Bengal, Calcutta, 1930, p. 23.

DOBBS, A. C-' Green-Manuring in India,' Bull. 56, Agric. Research Institute, Pusa, 1916.

FOWLER, G. J.--'Recent Experiments on the Preparation of Organic Matter,' Agric. Journ. of India, 25, 1930, p 363.

HALL, A. D.--The Book of the Rothamsted Experiments, London, 1905.

HOWARD, A. and HOWARD, G. L. C.--'The Improvement of To-bacco Cultivation in Bihar,' Bull. 50, Agric. Research Institute, Pusa, 1915.

HOWARD, A.--Crop Production in India, a Critical Survey of its Problems, Oxford University Press, 1924.

HOWARD, A. and HOWARD, G. L. C.--The Application of Science to Crop Production, an Experiment carried out at the Institute of Plant Industry, Indore, Oxford University Press, 1929.

HUTCHINSON, H. B. and RICHARDS, E. H.--'Artificial Farmyard Manure,' Journ. of the Min. of Agric. (London), 28, 1921, p. 398.

KING, F. H.--Farmers of Forty Centuries, or Permanent Agriculture in China, Korea and Japan, London, 1926.

LOHNIS, F.--'Nitrogen Availability of Green Manures,' Soil Science, 22, 1926, p. 253.

LOHNIS, F.--'Effect of Growing Legumes upon succeeding Crops,' Soil Science, 22, 1926, p. 355.

RUSSELL, E. J.--Soil Conditions and Plant Growth, London, 1927.

RUSSELL, E. J.--'The Present Status of Soil Microbiology,' Proc. of the Inter. Congress on Soil Science, Washington, D.C., 1928, p. 36.

RUSSELL, E. J. and RICHARDS, E. H.--'The Changes taking place during the Storage of Farmyard Manure,' Journ. of Agric. Science, 8, 1927, p. 495.

RUSSELL, E. J. and others.--The Micro-organisms in the Soil, London, 1923.

WAKSMAN, S. A.--'Chemical and Microbiological Principles underlying the Decomposition of Green-Manures in the Soil,' Journ. of the Amer. Soc. of Agronomy, 21, 1929, p. 1.

WAKSMAN, S. A., TENNEY, F. G. and DIEHM, R. A.--'Chemical and Microbiological Principles underlying the Transformation of Organic Matter in the Preparation of Artificial Manures,' Journ. of the Amer. Soc. of Agronomy, 21, 1929, p.533.

WAKSMAN, S. A. and DIEHM, R. A.--Chemical and Microbiological Principles underlying the Transformation of Organic Matter in Stable Manure in the Soil,' Journ. Of the Amer. Soc. of Agronomy, 21, 1929, p.795.

IV. THE MANUFACTURE OF COMPOST BY THE INDORE METHOD

The aim of the Indore method of manufacturing compost is by means of a simple process to unite the advantages of three very different things: (1) the results of scientific research on the transformation of plant residues; (2) the agricultural experience of the past, and (3) the ideal line of advance in the soil management of the future--in such a manner that all the by-products of agriculture can be systematically converted into humus. An essential feature of this synthesis is the avoidance of anything in the nature of fragmentation of the factors. All available vegetable matter, including the soiled bedding from the cattle-shed, all unconsumed crop residues, fallen leaves and other forest wastes, farmyard manure, green-manures and weeds pass systematically through the compost factory, which also utilizes the urine earth from the floor of the cattle-shed together with the available supply of wood ashes from the blacksmith's shop and the workmen's quarters. The only other materials employed are air and water. This manufacture is continuous right through the year, including the rainy season, when a slight modification has to be made to ensure sufficient aeration. The product is a finely divided leaf-mould, of high nitrifying power, ready for immediate use. The fine state of division enables the compost to be rapidly incorporated and to exert its maximum influence on a very large area of the internal surface of the soil.

The Indore process thus utilizes all the by-products of agriculture and produces an essential manure. Besides doing this any successful system of manufacturing compost must also fulfil the following conditions:--

1. The labour required must be reduced to a minimum. The process must fit in with the care of the work cattle and with the ordinary working of the farm.

2. A suitable and also a regular carbon-nitrogen ratio must be produced by well mixing the vegetable residues before going into the compost pits. Unless this is arranged for, decay is always retarded. The mixing of these residues, combined with the proper breaking up of all refractory materials is essential for rapid and vigorous fermentation and for uniformity throughout the process.

3. The process must be rapid. To achieve this it must be aerobic throughout, and must include arrangements for an adequate supply of water and for inoculation, at the right moment, with the proper fungi and bacteria. The general reaction of the mass must be maintained, within the optimum range, by means of earth and wood ashes. The maintenance of the proper relationship between air and water, so that no delay takes place in the manufacture, proved to be the greatest practical difficulty when evolving the process.

4. There should be no losses of nitrogen at any stage; if possible, matters should be so arranged that fixation takes place in the compost factory itself and afterwards in the field. To conserve the nitrogen, the manufacture must stop as soon as the compost reaches the nitrification stage, when it must either be used or banked. It can best be used as a top dressing for irrigated crops; it can be preserved, as money is kept in a bank, by applying it to the fields when dilution with the large volume of soil arrests further changes till the next rains.

5. There must be no serious competition between the last stages of the decay of the compost and the work of the soil in growing a crop. This is accomplished by carrying the manufacture of humus up to the point when nitrification is about to begin. In this way the Chinese principle of dividing the growing of a crop into two separate processes--(1) the preparation of the food materials outside the field, and (2) the actual growing of the crop--can be introduced into general agricultural practice.

6. The compost should not only add to the store of organic matter and provide combined nitrogen for the soil solution but should also stimulate the micro-organisms.

7. The manufacture must be a cleanly and a sanitary process from the point of view both of man and also of his crops. There must be no smell at any stage, flies must not breed in the compost pits or in the earth under the work cattle. The seeds of weeds, the spores of harmful fungi, the eggs of noxious insects must first be destroyed and then utilized as raw material for more compost. All this is achieved by the combination, in the compost pits during long periods, of high temperature and high humidity with adequate aeration.

THE COMPOST FACTORY

The compost factory at Indore adjoins the cattle shed. This latter has been constructed for forty oxen and is provided with a cubicle, in which a supply of powdered urine earth can conveniently be stored. The cattle stand on earth. A paved floor is undesirable as the animals rest better, are more comfortable and are warmer on an earthen floor. The earth on which the cattle stand absorbs the urine, and is replaced by new earth to a depth of six inches every three or four months. The compost factory (Plates III and IV show the cattle shed and compost factory) itself is a very simple arrangement. It consists of thirty-three pits, each 30 ft. by 14 ft. and 2 ft. deep with sloping sides, arranged in three rows with sufficient space between the lines of pits for the easy passage of loaded carts. The pits themselves are in pairs, with a space 12 ft. wide between each pair. This arrangement enables carts to be brought up to any particular pit. Ample access from the compost factory to the main roads is also necessary, so that during the carting of the compost to the fields, loaded and empty carts can easily pass one another, and also leave room for the standing carts which are being filled. For a large factory it is an advantage to have water laid on, so that the periodical moistening of the compost can be done by means of a hose pipe. At Indore, water is pumped through a 3 in. pipe into a pressed steel tank, 8 ft. by 8 ft. by 8 ft., holding 3,200 gallons, which is carried on walls, 4 ft. above the ground, to provide the necessary head. This supply lasts about a week. Water is led by 1-1/2 in. pipes from the tank to eight taps, to which the armoured hose can be screwed. Each tap serves about six pits. The general arrangement will be clear from Plate IV.

THE WASTE PRODUCTS OF AGRICULTURE

The total cost of the water tank, including arrangements for distribution, was Rs. 1650 (equivalent to about 120 pounds sterling). This was made up as follows: tank, Rs. 750; pipe system, Rs. 466; girders for tank, Rs. 31; armoured hose, Rs. 28: railway freight, Rs. 88; masonry work, Rs. 148; labour, including fitting up, Rs. 129.

The space under the tank, which is walled in on three sides and is open on the leeward side, is used for storing wood ashes, and for keeping the tubs and implements needed for the making of compost.

For a smaller factory or for the small holder, such a water system is not necessary. All that is needed is that the compost pits should be arranged near a well.

COLLECTION AND STORAGE OF THE RAW MATERIAL

Plant Residues.--All vegetable wastes from the cultivated area--such as weeds, cotton and other stalks, green-manure, cane-trash, fallen leaves and so forth, and all inedible crop residues from the threshing floor--are carefully collected. All woody materials like cotton and pigeon-pea (Cajanus Indicus Spreng.) stalks are crushed by placing on the farm roads to be trampled and reduced by the traffic to a condition resembling broken up wheat straw (Plate V). All green materials--such as weeds and green-manures--are withered for at least two days before use or storage. All these various residues are stacked near the cattle shed as received, layer by layer--if possible under cover during the rains--so that these materials may become thoroughly mixed. Each layer must not be more than one foot thick, otherwise difficulties arise in making a suitable mixture. Care must also be taken to remove the stacked material in vertical slices so as to ensure even mixture. Very hard and woody materials--such as sugarcane and millet stumps, wood shavings, sawdust and waste paper--should be dumped separately in one of the empty compost pits with a little earth and kept moist. After this preliminary treatment, these hard and resistant materials can be readily composted. Steeping such materials in water for two days, before addition to the bedding under the work cattle, serves the same purpose.

Urine Earth and Wood Ashes.--All the earth removed from the silage pits, all earthy sweepings from the threshing floors and all silt from drains are stored in a convenient place near the cattle-shed. This provides an adequate supply of suitable earth for absorbing the urine of the work cattle, and acting as a base in the making of compost. This earth is spread evenly on the cattle-shed floor to a depth of six inches and renewed every three or four months. Half the urine earth when removed from the floor should be crushed (Plate VI) in a mortar mill(See Plate's V and VI). to break up the large lumps, and should be stored under cover as dry powdered urine earth. The other half of the urine earth should be applied direct to the fields as manure. All available wood ashes should be stored under cover, as in the case of the powdered urine earth. These materials (urine earth and wood ashes) are as essential in the manufacture of compost as the plant residues themselves.

Water and-Air. Both water and air are needed for the compost process, which therefore must be carried out near a well or other source of fresh water.

ARRANGEMENT AND DISPOSAL OF THE BEDDING UNDER THE WORK CATTLE

(All quantities in the following refer to one pair of oxen. The figures should be multiplied, when necessary, by the number of pairs of oxen kept.)

All the uneaten food and any damaged silage are thrown on the wet portions of the cattle-shed floor. One and a half pals of stacked vegetable refuse, together with not more than one-twentieth of this amount of hard resistant material (such as wood shavings, sawdust or waste paper) from the soaking pit are spread on the floor. (A pal is a stretcher made of a piece of gunny sheet (4 ft. by 3 ft.) nailed to two bamboos each 7 ft. 6 in. long.) The cattle sleep on this bedding during the night. In this way the bedding gets crushed and broken still further and also impregnated with urine. Next morning one-fourth of a tagari of fresh dung is removed to the compost pit; the rest of the cattle dung being scattered on the bedding in lumps not bigger than a small orange; or this excess dung can be made into cow-dung cakes (kundas) for fuel. (A tagari is a bowl made of sheet iron, capacity five-sevenths of a cubic foot. In Table IV the metal bowls are converted into pounds or double handfuls of the materials used. Kundas, thin flat cow-dung cakes, about twelve inches in diameter and one inch thick, are used in the villages of India as fuel for the cooking of food.) Two-fifths of a tagari of dry urine earth is sprinkled on the used bedding in the same manner as murum (Murum is the Hindustani name of the permeable layer of decayed basalt which underlies the black cotton soils of India) is spread on roads. The bedding is then transferred by a spade on to the pal from one end to the other and removed to the compost pit. In this way the raw material used for the compost is made perfectly homogeneous. The earthen floor of the cattle-shed should then be swept clean, the sweepings being removed on a pal to the compost pit. All wet patches on the floor are covered with new earth, after scraping out the very wet portions. In this way all smell in the cattle-shed is avoided and the breeding of flies in the earth underneath the animals is entirely prevented. Bedding for the next day can then be laid as described above.

During the rains, the bedding should consist of three layers--a bottom layer and a top layer of dry material specially reserved for the purpose, any withered residues being sandwiched in between. On very wet days, all the urine earth may be added to the bedding before removal to the compost heap.

The volume and weight of the various materials which are moved to and fro in the sheet-iron bowls (tagaries) are given in Table IV.

TABLE IV

VOLUME (IN DOUBLE HANDFULLS) AND WEIGHT (IN LB.) OF THE CONTENTS OF A TAGARI

	Volume in double handfuls	Weight in lb.
Fresh dung	6.5	39.5
Powdered urine earth	20.5	22.5
Wood ashes	15	20
Fungus innoculant	6	20
Bacterial innoculant	6	20
Refractory vegetable residues		9
Mixed vegetable residues		9
Impregnated bedding		16.5
Sweepings from the cattles-shed floor		19

CHARGING THE COMPOST PITS

A convenient size for a compost pit is 30 ft. by I4 ft. and 2 ft. deep with sloping sides. The depth of the compost pit is most important on account of the aeration factor. It should never exceed 24 in. A wooden tub, a rake, a bowl (tagari), and a few empty kerosine tins (each holding four gallons) with handles are all that are needed besides the pal.

The following materials are placed alongside each compost pit--powdered urine earth, two fifths tagari, fresh dung, one-quarter tagari; fungus material, three tenths tagari, taken from a compost pit ten to fifteen days old; wood ashes, one twentieth tagari; water, one kerosine tin. The wood ashes and one twentieth of a tagari of urine earth are mixed with some dung and fungus material in a portion of the water to make a thin slurry. The pals of bedding should be added, as they arrive, from one edge of the pit by simply allowing the bamboo pole of the pal next the pit to fall into it (Plate VII). The other pole is then lifted so that the rest of the bedding drops easily into the pit. The material is then spread by means of the rake in a layer, not exceeding two inches thick over the compost pit. All trampling of the charged pit must be avoided as this interferes with aeration. Some dry urine earth and then the stirred slurry are first sprinkled thinly on each charge of bedding, which should appear evenly wetted. The soaked residues from the tub are then scattered on each layer of bedding. This inoculates the mass with active fungus throughout. The polished surfaces of the bedding are also covered with an active adherent coating. This leads to rapid and even crumbling. The volume of the slurry is made up with more dung, fungus starter and water as required. The pit is charged with the bedding, layer by layer, until all the bedding is used up. The sweepings from the cattle-shed floor, which are rich in urine, are sprinkled on the top of each day's charge with a tagari, followed by one-third of a tin of fresh water. This distributes the urine evenly throughout the daily charge and also prevents excess drying. Another watering in the evening, with two-thirds of a tin, and a third watering the next morning with one-third of a tin completes the charge. The pit or a suitable portion of it should be filled up to the brim in six days or less, the remaining part being filled subsequently.

The period of charging must not exceed six days, whether or no the pit is completely filled by then. Each six days' charge should be regarded as one unit in the manufacture of compost, no matter whether the pit is filled completely or not.

Everything is now ready for the development of an active fungus growth (the first stage in the manufacture of compost). When properly managed, a vertical section of the fermenting mass should appear quite uniform and should not show any alternate layers.

As the pits are frequently full of water during the greater part of the rains, the compost must be made in heaps from the middle of June to I October. The dimensions of the heaps should not exceed 7 ft. by 7 ft. at the top, 8 ft. by 8 ft. at the bottom and 2 ft. in height. The dimensions of these monsoon heaps (any one of which is not necessarily completed by the amount of vegetable waste which can be accumulated in six days) must not be exceeded, otherwise aeration difficulties are certain to be encountered. The decomposition in heaps during the rains does not take place so evenly as in the pits.

THE WASTE PRODUCTS OF AGRICULTURE

During the early rains, all the material in the pits must be transferred to heaps on the surface. This is most conveniently done at the time of the first, second or third turn.

The subsequent waterings are most important, otherwise decay will stop. The first watering is done twelve days (counted from the date on which the filling of the pit begins) after charging, when 1.25 tins are added evenly over the whole surface. Further water is added at the time of the first, second and third turning and afterwards as needed. During the rains, the quantity of water as given above must be added at the time of charging; the subsequent waterings during the rains may be reduced or completely omitted according to the weather. Stagnant rain-water from the pits should never be used. When watering is done by a hose pipe from a tank as at Indore, the amount added can easily be adjusted if the rate of flow is known.

TURNING THE COMPOST

To ensure uniform mixture and decay, and to provide the necessary amount of water and air as well as a supply of suitable bacteria, it is necessary to turn the material three times. The only difficulty which is likely to arise in the process is the establishment of anaerobic conditions between the period of charging and the first turn. This can be caused by over watering or by want of attention to the mixing. It is at once indicated by the smell and by the appearance of flies attempting to breed in the mass. When this occurs, the heap should be turned at once with the addition of dung slurry and wood ashes.

First turn. Sixteen days after charge (Plate IX). Sufficient fresh water should be ready--about four tins according to the season. Three-fifths of a tagari of compost is taken from another pit thirty days old (just after the second turn) and scattered on the surface of the material. This is necessary for inoculating the mass with the proper bacteria. The top layer of the compost is then loosened and mixed, a portion at a time, with a rake and well moistened with water. Half the heap is sliced with a spade a few inches breadth wise and vertically from top to bottom to fill one tagari at a time. Tagari after tagari is poured in rows on the other undisturbed half to make a layer which is then sprinkled with water. This is repeated until one-half of the contents of the pit is doubled lengthwise over the other. The heap is then watered, sufficient being added at this first turn to prevent the wasteful use of water afterwards. After turning, the heap should not rise more than twelve inches above ground level. The second watering, 1.5 tins, is given twenty-four days after charge. At the first turn, the materials should be arranged on the windward side of the pit to avoid the cooling of the mass and also excessive drying. During the rains, when heaps are made, it is not possible to double one-half of the heap over the other. The material should then be completely turned and the heap re-made. The heaps should be made as near as possible to each other.

Second turn. One month after charge (Plate IX). The water required is about three tins. The material is cut vertically in two inch slices and piled up with watering as before along the empty half of the pit. The material should fall loosely, under each stroke of the spade and not in lumps, so as to ensure copious aeration. The third and fourth waterings, 1.5 tins each, are given five and six weeks after charge.

Third turn. Two months after charge. About two tins of water are necessary. A rectangular heap is made on the ground alongside the pit or in the field not more than 10 ft. broad at the base, 9 ft. wide at the top and 3.5 ft. high, the material being spaded and piled with watering as before. When the heap is made in the field, all the water needed should be added at the time of carting. The contents of several pits may now be placed side by side to save space, to economize water and to facilitate removal. The fifth and sixth waterings, 1.25 tins each, are given nine and ten weeks after charge. For the first time during the process, extra labour, namely three men and four women for six hours, is required for each pit at the third turn. As the heap can be made either in the factory or in the field, this additional labour can be debited to the application of the humus to the land.

Three months after charge the manure is ready, when it should be applied to the land. If kept in heaps longer than three months after charge, nitrogen is certain to be lost. There is no great harm in putting the manure on the land after two months if urgently required, especially when the process has run for some time and everything is in full working order.

THE WASTE PRODUCTS OF AGRICULTURE

TIME-TABLE OF OPERATIONS

The complete time-table of the manufacture of compost, which takes ninety days, is given in Table V.

TABLE V

THE COMPLETE TIME-TABLE FOR ONE COMPOST PIT

Day	Event
1	Charging begins.
6	Charging ends.
10	Fungus growth established.
12	First watering.
16/17	First turning, compost inoculated with bacteria from another pit thirty-days old.
24	Second watering.
30/32	Second turning.
38	Third watering.
45	Fourth watering.
60	Third turning.
67	Fifth watering
75	Sixth watering
90	Removal to field.

OUTPUT

Fifty cart-loads of ripe compost per pair of oxen per annum can be made from the plant residues available on any holding. The quantity can be more than doubled when all the dung and urine earth are used, provided of course sufficient vegetable refuse can be secured. Fifty to seventy-five tins (200 to 300 gallons) of water, according to the season, are sufficient to make one cart-load of finished compost. No extra labour is required other than that usually employed in the cattle-shed, namely two men and three women. These are sufficient for the work connected with forty oxen and the preparation of 1,000 carts of compost per annum.

The labour needed for the annual manufacture of 1OOO cart-loads of compost has been reduced to a minimum by: (1) the provision of a water supply; (2) the general design of the cattle-shed and compost factory and (3) the detailed training of the labour force to carry out the work quickly and without unnecessary fatigue. This aspect of the manufacture of humus has been greatly assisted by the system of managing labour adopted at the Institute (Appendix D).

During the year 1930, when 840 cart-loads of compost were prepared, a careful record of the actual time spent on compost making by the labour employed to look after the work cattle, was made. It was found that one half of the time of this labour was spent on the care of the cattle and one half on the making of compost. The total wages debited to actual compost making came to Rs 441.5, i.e. to 8.5 annas, or nine pence halfpenny, per cart-load of finished material. During the present year, 1931, the output has increased and is expected to reach 1,000 cart- loads. It is best to spread the compost on the land directly it becomes ready, so as to facilitate the distribution of farm work throughout the year.

MANURIAL VALUE OF INDORE COMPOST

One-cart load of Indore compost is equivalent, as regards nitrogen content, to two cart-loads of ordinary farmyard manure. Properly made compost has another great advantage over ordinary manure, namely its fine powdery character which enables it to be uniformly incorporated with the soil and to be rapidly converted into food materials for the crop. Taking everything into consideration, Indore compost has about three times the value of ordinary manure.

V. THE CHIEF FACTORS IN THE INDORE PROCESS

The Indore process enables the Indian cultivator to transform his mixed vegetable wastes into humus; in other words to become a chemical manufacturer. The reactions involved are those which take place under aerobic conditions during the natural decay of organic residues in the soil. The object of the process is to bring these changes under strict control and then to intensify them. A knowledge of the chemical processes involved and of their relative importance is therefore essential in applying the process to other conditions. These matters form the subject of the present chapter.

THE CONTINUOUS SUPPLY OF MIXED VEGETABLE WASTES

A continuous supply of mixed vegetable wastes throughout the year, in a proper state of division, is the chief factor in the process. The ideal chemical composition of these materials should be such that, after the bedding stage, the carbon-nitrogen ratio is in the neighbourhood of 33:1. The material should also be in such a physical condition that the fungi and bacteria can obtain ready access to, and break down the tissues without delay. The bark, which is the natural protection of the celluloses and lignins against the inroads of fungi and bacteria, must first be destroyed. This is the reason why all woody materials-- such as cotton-stalks, pigeon-pea stalks and sann hemp (Crotalaria juncea L.)--are laid on the roads and crushed by the traffic into a fine state of division before composting. Still more refractory residues like the stumps of sugar-cane and millets, shavings, sawdust, waste paper and packing materials, old gunny bags and similar substances, must either be steeped in water for forty-eight hours or mixed with moist earth in a pit for a few days before passing, in small quantities daily, into the bedding.

The vegetable wastes which have been utilized at Indore for the last six years are the following:--

Residues available in large quantities: Cotton stalks, sann hemp--either as green plants reaped before the flowering stage or as dried stems of the crop kept for seed, pigeon-pea stalks, sugar-cane trash, weeds, fallen leaves.

Residues available in moderate quantities: Mixed dried grass, gram stalks, wheat straw, uneaten and decayed silage, millet stalks damaged by rain, residues of the safflower crop, ground-nut husks, ground-nut stalks and leaves damaged by rain, sugar-cane and millet stumps.

Residues available in small quantities: Waste paper and packing materials, shavings, sawdust, worn out gunny bags, old canvas, worn out uniforms, old leather belting.

. . . . The raw materials available at Indore differ greatly in chemical composition and particularly in the percentage of nitrogen. Many of these wastes, such as cotton-stalks, the stems of sann hemp and of the pigeon-pea, and cane trash are too low in nitrogen for rapid composting. Others--such as green hemp, reaped just before flowering, ground-nut residues and leguminous and other weeds--contain higher percentages of nitrogen, a portion of which is certain to be lost during the process if these materials are composted singly. A proper mixture of the various materials available, so that the nitrogen content of the mass throughout the year is kept uniform and sufficiently high, is the first condition of success. For this reason it is necessary to collect and stack the various residues in such a manner that a regular supply of dry, mixed, vegetable wastes (as already stated with a carbon-nitrogen ratio in the neighbourhood of 33:1 after the material has been used as bedding) is available right through the year. This could only be accomplished at Indore: (1) by cutting the cotton-stalks soon after picking is over so as to secure the maximum number of leaves; (2) by growing a large area of sann hemp, which contains when withered as much as 2.3 per cent of nitrogen; and (3) by securing as much green weeds, groundnut residues and fallen leaves as possible for the mixture. All these materials are rich in nitrogen, and help to bring the carbon-nitrogen ratio near the required standard. By stacking the various constituents in layers, not more than one foot thick, and by a judicious admixture with the residues richest in nitrogen, it is possible to provide a continuous supply of dry mixed material of the correct chemical composition. During the rains, a good deal of the raw material is in the form of fresh green weeds, rich in nitrogen and soluble carbo-hydrates. These must be spread, in thin layers, on the grass borders of the fields alongside the roads and withered, before being carried to the stack or used as one of the constituents of the bedding. Only in this way can the most be made of this valuable material. Collecting weeds in temporary heaps on the borders of fields leads to serious waste of the soluble carbo-hydrates and also of the nitrogen.

COMPOSTING SINGLE MATERIALS

A number of experiments have been carried out at Indore during the last four years with the following single materials--cotton-stalks, pigeon pea stalks, cane trash, weeds (green and withered), sann hemp (green and withered). When necessary these residues were either passed through a chaff cutter or crushed with a disc harrow before composting direct in heaps, eighteen inches high, or in pits filled to the same depth. In some cases Adco was employed as the source of nitrogen and base, in others cattle-dung and urine earth were used. Sufficient water was always added to maintain a high moisture content.

Although the cotton residues, fermented direct with urine earth and cattle-dung, contained 16.5 per cent of green leaves (high in nitrogen) and every care was taken to maintain the correct relation between air and water, the results were not completely satisfactory. Fermentation was rapid at the beginning, due to the presence of the leaves, but slowed down afterwards. It took 150 days to obtain a usable product, as compared with the ninety days required for mixed wastes.

In the case of cotton-stalks, broken down by the use of Adco, the results were still more unsatisfactory. Several interesting facts however came to light. The fermentation tended to be uneven; the temperature of the heaps was always irregular; the mass did not retain moisture well; a very large quantity of water was needed. The final product, although high in nitrogen, tended to be somewhat coarse and to contain a good deal of partially decomposed material. The maximum temperatures in the Adco heaps during the first 100 days fell from 53.5 degrees C. to 29.5 degrees C. (In the standard Indore process, the range of temperature during ninety days was 65 degrees C. to 33 degrees C.) The final product was fairly satisfactory as regards fineness (80.5 per cent passed through a sieve of six meshes to the linear inch) and high in total and available nitrogen (total 2.50, available 0.42 per cent). The corresponding figures for the product made from cotton-stalks with cattle-dung and urine earth were--fineness 84.2 per cent and total nitrogen 1.61, of which 0.I3 per cent was available. In spite of the higher nitrogen content obtained in the Adco product, no increase in growth was obtained when equal quantities of both kinds of cotton-stalk compost were used in pot cultures of millet. This result probably follows from the fact that the use of Adco often produces compost with a carbon-nitrogen ratio narrower than 10:1, the ideal which should be aimed at in the manufacture of humus. The extra nitrogen in such cases is always liable to be lost before the crop can make use of it.

The results obtained in the direct composting of other single materials, like pigeon-pea stalks and cane trash, were still more unsatisfactory. When used alone, either with cow-dung and urine earth or with Adco, little change took place in a month in spite of copious watering and occasional stirring. When, however, these materials were passed through the cattle-shed and used as bedding, the results were distinctly better but not really satisfactory. At the end of six months, the heaps were only about half decomposed.

Difficulties also arise when weeds (fresh or withered) or sann hemp (fresh or withered) are composted by themselves or when a mixture of the two is employed. In the first place, the nitrogen content of this material is too high and serious losses of this element take place. In the second place, these residues, particularly when fresh, tend to pack closely in the heaps and to prevent sufficient aeration (Table VIII). For this reason, withered weeds or withered sann must never form more than about 30 per cent of the volume of the bedding, the rest being made up of mixed residues like cotton and pigeon-pea stalks with a much lower nitrogen content.

TABLE VIII

LOSSES OF NITROGEN RESULTING FROM THE CLOSE TEXTURE OF THE MASS

No. of pit	Withered materials used	Total nitrogen (lb.) at beginning	Total nitrogen (lb.) in the finished product	Loss or gain of nitrogen (lb.)	Percentage loss or gain
34	Weeds	44.2	25.7	−18.5	−41.8
38	Half sann, half weeds	42.8	28.4	−14.5	−33.8
40	"do."	49.7	29.2	−20.5	−41.3
41	Mixed residues	28.3	29.5	+1.3	+4.4

When one food material at a time is provided for the fungi and bacteria, loss of nitrogen or aeration difficulties or both always occur. When a mixed diet is employed, everything goes smoothly, provided of course all other important details receive attention.

NITROGEN REQUIREMENTS

The total amount of combined nitrogen which must be added to the mixed residues for the use of the micro-organisms is less than was at first expected. The vegetable wastes from the 300 acres of land at the disposal of the Institute can be converted into humus by means of half the urine earth and one quarter of the cattle-dung of the forty oxen maintained for the work of the station. A satisfactory product, with a suitable carbon-nitrogen ratio, can be obtained with this reduced supply of dung (Table IX). At first all the urine earth was employed in composting, but it was soon found that better aeration resulted with only half the quantity. Although in many cases the compost made with full dung contains about 0.15 per cent more nitrogen than that made with reduced dung, the results obtained in the field were always the same. The surplus urine earth is used for manuring the land, the extra cattle-dung can either be used up in composting or can be sold for the manufacture of cow-dung cakes (kundas). This means: (1) that the present high output of compost could be doubled if sufficient vegetable wastes could be obtained; and (2) that even after this increased output is reached, half the dung would still be in excess.

TABLE IX

RESULTS WITH REDUCED (ONE-FOURTH) AND FULL DUNG

No. of pit	Amount of dung used	Total N (lb.) at begin	Total N (lb.) at end	Percent gain of N	Percent of N at begin	Percent of N at end	Carbon nitrogen ratio	Fineness
14	Reduced	29.12	32.36	11.1	0.67	0.84	11.6:1	88.5
15	Full	32.70	34.87	6.6	0.70	0.72	12.6:1	82.5

The fact that the cultivator really requires only a fraction of his cow-dung for converting all his vegetable wastes into humus, disposes once and for all of the view that the salvation of Indian agriculture lies in substituting some other fuel for cow-dung cakes. This material is essential for the slow cooking needed for a vegetarian diet As no other suitable fuel. exists in many of the villages of India, cow-dung must be utilized. Fortunately, when all the available vegetable wastes have been converted into humus, a large supply of cow-dung for fuel will still be available, and there is no reason why it should not be burnt. The ashes, however, should be carefully collected and employed as a base in the compost process.

In all the comparative trials which have been made at Indore, with Adco on the one hand and with urine earth and cow-dung on the other hand, far more satisfactory results have been obtained with the indigenous materials. The weak point of Adco is that it does nothing to overcome one of the great difficulties in composting, namely the absorption of moisture in the early stages. In the hot weather in India, the Adco pits lose moisture so rapidly that the fermentation stops, the temperature becomes uneven and then falls. When, however, urine earth and cow-dung are used, the residues become covered with a thin colloidal film, which not only retains moisture but contains the combined nitrogen and minerals required by the fungi. This film enables the moisture to penetrate the mass and helps the fungi to establish themselves. Another disadvantage of Adco is that when this material is used according to the directions, the carbon-nitrogen ratio of the final product is narrower than the ideal 10:1. Nitrogen is almost certain to be lost before the crop can make use of it, particularly when Adco compost is added to the land some weeks before sowing takes place.

THE AMOUNT OF WATER NEEDED

It is an easy matter to waste large quantities of water in the process. As a result of repeated trials, the maximum economy of water is obtained when 168 gallons (for every 400 tagaries of used bedding) are added at the time of charging and during the next twenty-four hours. After this, the watering should proceed as laid down in Chapter IV. Any departure, in either direction, leads to a waste of water. . . The standard water requirements as now adopted, per cart-load of finished compost, varies from 200 to 300 gallons according to the season. The Malwa Plateau, on which Indore is situated, is a windswept area in which the humidity is low for at least eight months in the year. It is unlikely, therefore, that these quantities will be greatly exceeded, except in very dry areas like the Punjab and Sind.

At the beginning of the process, care should be taken to add just sufficient moisture to keep the average water content below 50 per cent of complete saturation, so as to help the fungi to establish themselves rapidly and strongly. This matter is important, as the vegetable wastes take up water very slowly at the beginning. If too much is added at this stage, free water tends to accumulate in the air spaces and to hinder aeration. This checks the growth of the fungi, which thrive best if the total moisture is below 50 per cent. The moment the crumbling of the material sets in, water is absorbed more rapidly. After the first turn and till the compost is ready to cart to the fields, the total moisture content should vary between 50 and 60 per cent. After the final turn, when no more water is added, the percentage again drops to what it was at the beginning, namely under 50 per cent. During the rains, the water content of the heaps naturally tends to run a little higher than in the dry season. The depressing effect on the fermentation of very heavy monsoon downpours was well brought out during a wet period of seven days (10-16 September 1930), when 12.86 inches occurred, 11.65 inches of which were received in one continuous fall, lasting seventy-two hours. At the end of this spell, there was a temporary fall in the temperature of the heaps. Three or four days after the downpour stopped, fermentation again became vigorous as is seen by the rapid rise in the temperature (Table XII).

TABLE XII

THE EFFECT OF HEAVY RAINFALL (12.86 INCHES) ON THE TEMPERATURE OF FERMENTING HEAPS

(temperature in degrees C)

Age of Heap	Before the rain	After the rain	Three days after the rain stopped
First week	61	44	53
First week	59	37	50
After the first turn	55	38	52
After the first turn	54	39	53
After the second turn	51	30	48
After the second turn	48	32	48
After the third turn	41	29	38
After the third turn	40	30	38

THE SUPPLY OF AIR

The control of the aeration factor is perhaps the most difficult part of the process, and requires careful attention. The first condition of success in obtaining a sufficient supply of oxygen and nitrogen for the micro-organisms, is the use of mixed bedding which maintains an open texture through out the process. As already explained, single materials always tend to pack too closely and to cut off the air supply. The second condition of success is attention to detail at the time of charging. The bedding must be carefully spread, the urine earth, the cow-dung slurry and the wood ashes must be evenly scattered. Water must be properly distributed over the whole mass, and there must be no trampling. At the time of the first and second turns, the spading or forking must be carried out so that the material falls lightly, when thorough mixing takes place with the maximum amount of aeration. The third condition of success concerns the depth of the pit or heap, which must never exceed twenty-four inches. This is the maximum distance to which air can penetrate the fermenting mass in sufficient volume. If this depth is exceeded, two things happen: (1) the decay of the layers below twenty-four inches is retarded; (2) is always a loss of nitrogen through denitrification (Table XIII).

TABLE XIII
COMPOST MAKING IN DEEP AND SHALLOW PITS

	Pits 4 ft. deep	Pits 2 ft. deep
Amount of material (lb.) in charge	4500	4514
Total nitrogen (lb.) at the beginning	31.25	29.12
Total nitrogen (lb.) at the end	29.49	32.36
Loss or gain of nitrogen (lb.)	-1.76	+3.24
Percentage loss or gain of nitrogen	-6.1	+11.1

The air supply can also be permanently interfered with if too much earth and cow-dung are used at the time of the first charge. These materials make the whole mass too solid and pack it too closely. Anaerobic conditions are then established. This is indicated by the smell and by the appearance of flies, which then find suitable breeding conditions. The remedy is at once to turn the material, with the addition of cow-dung slurry and wood ashes. Temporary interruptions in aeration also follow over watering or the soaking due to heavy rain. These troubles, however, pass in two or three days as the heap dries and the surplus moisture is gradually taken up by the mass (Table XII).

THE MAINTENANCE OF THE GENERAL REACTION

In order to maintain the general reaction of the mass within the optimum range, a suitable base is necessary for neutralizing excessive acidity, and for the temporary absorption of any ammonia that may be given off during the process. The urine earth and wood ashes provide this in the most economical manner. Black cotton soil (Table XIV) contains an ample reserve of weak bases. The buffering effect of these maintains the general reaction constant throughout (Table XV). Further, black soil contains a high percentage of clay, the colloids of which are most useful in two ways.

TABLE XIV
MECHANICAL AND CHEMICAL ANALYSES OF BLACK SOIL

Mechanical

Fraction	I	II	III
Clay	42.5	45.6	38.3
Fine Silt	19.6	21.8	17.7
Silt	12.5	10.8	11.3
Fine Sand	7.4	4.2	6.7
Coarse Sand	10.2	6.0	3.0
Moisture	3.3	6.4	3.0
Loss on Ignition	3.0	5.7	2.7
Calcium Carbonate	1.6	6.1	1.4

Chemical
Constituent

Insolubles	56.1	73.8	68.7
Fe_2O_3	9.8	9.1	11.2
MnO_2	--	0.1	0.3
CaO	6.6	0.9	1.0
MgO	2.5	1.5	1.8
K_2O	0.4	0.2	0.4
Na_2O	0.2		
P_2O_5	0.08	0.17	0.06
CO_2	0.8	0.1	0.4
N	0.03	0.05	0.05
Organic combined water	9.4	7.4	5.83

TABLE XV
REACTION AND TEMPERATURE IN THE INDORE PROCESS

Stage	pH value	Temperature in degrees C.
One day after charge	7.2	63
After first turn (19 days old)	7.4	49
After second turn (34 days old)	7.5	45
After third turn (60 days old)	7.6	41
Ripe manure (90 days old)	7.7	35

In the first place, these substances are capable of temporarily absorbing, till required for oxidation, any ammonia given off in the process. In the second place, the colloids, when mixed with the urine and cow-dung, cover the vegetable wastes with a thin, nutrient, moisture-retaining film which is of the utmost value, not only in the gradual absorption of water but also in providing the fungi with a favourable nidus for the steady breaking down of the vegetable wastes. The result is the rapid establishment of a vigorous mycelial growth, and the early crumbling of the whole mass. When a colloidal film is not employed, as in the Adco process, it is most difficult to get the material to absorb and retain sufficient moisture. Consequently, an even and vigorous mycelial growth is never quickly obtained. The colloids in soil are essential, both for coaxing water into the material and also for enabling the fungi to establish themselves rapidly and vigorously. The fungi are the storm troops of the composting process, and must be furnished with all the armament they need.

In a recent paper, received just as this chapter was completed, Jensen has shown that cellulose decomposing bacteria multiply most strongly at pH 7.0-8.0.

THE FERMENTATION PROCESSES

In addition to providing suitable conditions for the rapid development of the micro-organisms, it is necessary to inoculate the mass at the proper moment, so that there is no delay in the conversion. This is arranged for at the time of charging, when the pits are uniformly infected with actively growing fungus mycelium, taken from a compost pit ten to fifteen days old. At the same time, the bacteria present in cow-dung are introduced in large numbers. A further inoculation is carried out at the time of the first turn, when compost from a pit thirty days old is introduced into the mass. This provides a supply of the organisms required for the second half of the process.

The activity of the various micro-organisms can most easily be followed from the temperature records. A very high temperature, about 65 degrees C., is established at the outset, which continues for a long time with only a moderate downward gradient (Table XVI). This range fits in very well with the optimum temperature conditions required for the micro-organisms which break down cellulose. The aerobic thermophylic bacteria thrive best between 43 degrees and 63 degrees C.; the fungi between 40 degrees and 55 degrees C.

TABLE XVI
TEMPERATURE RANGE IN A NORMAL PIT

Moisture 45 to 55 per cent

Age in days	Temperature in degrees C
3	63
4	60
6	58
11	55
12	53
13	49
14	49
	First Turn
18	49
20	51
22	48
24	47
29	46
	Second Turn
37	49
38	45
40	40
43	39
57	39
	Third Turn
61	41
66	39
76	38
82	36
90	33

Period in days for each fall of 5 degrees C

Temperature Range (degrees C.)	No. of Days
65-60	4
60-55	7
55-50	1
50-45	25
45-40	2
40-35	44
35-30	14
Total	97 days

The temperatures throughout the fermenting mass are extraordinarily uniform in the pits; in the heaps the range is somewhat greater. An analysis of the figures shows that, before each turn, a definite slowing down in the fermentation takes place. As soon as the mass is remade, when more thorough admixture with copious aeration occurs, there is a renewal of activity during which the undecomposed portion of the vegetable matter from the outside of the pit or heap is attacked. At least three types of fermentation appear to be involved, which succeed one another with great rapidity. Two of these--those which occur between 50 degrees to 45 degrees C. and 40 degrees to 35 degrees C.--are long continued. It is during these latter stages of the process that the transformation of the vegetable wastes into humus occurs, accompanied by the rapid crumbling and shrinkage of the mass. A detailed analysis of the phases of micro-biological activity and the determination of the organisms concerned has not yet been carried out. The results, when obtained, cannot fail to throw considerable light on the real origin of humus and should also help to clear up a large field of rather obscure organic chemistry. This subject can naturally be more effectively studied in the mass under factory conditions than on a small scale in pot cultures or in the laboratory.

Wind is always a source of trouble and does most harm during the early stages of fermentation--between charging and the first turn--by lowering the temperature. The effect is most marked in the heaps which helps to explain why the process is not quite so efficient in the rains as it is in pits during the rest of the year.

The wind factor can be minimized during the rains by arranging the heaps so that they shelter each other. The pits must always be orientated so that the length is at right angles to the direction of the prevailing wind. This gives each pit a windward and a leeward side. The first turn must always be made towards the windward side, so that the earth wall of the pit protects the mass. Temporary spells of cold weather of short duration, such as occur in India, have no injurious effect. The fermentation is so vigorous that these sudden changes of temperature are not able to check the process. Hence in the tropics, compost houses are unnecessary. The disintegrating power of the process is so intense that unbroken stems of grass and weeds, several feet in length, are reduced in ninety days to partially decayed fragments only a few inches in length. The long continued moist heat of the fermentation also leads to other useful results besides helping to soften and break down the mass. The high temperatures make the process sanitary, and prevent all objectionable smell. Flies and other insects cannot breed in the hot mass. The seeds of weeds are killed in the process, as is shown by the fact that no weeds grow on the heaps of ripe compost. To confirm this point, 1s. of grass seeds were mixed with the bedding of two pits. Germination tests of the ripe manure gave negative results in each case.

GAINS AND LOSSES OF NITROGEN

A simple means of testing the efficiency of the process is to determine the amount of nitrogen lost. When vegetable wastes, with a carbon-nitrogen ratio in the neighbourhood of 33:1, are composted under strict aerobic conditions in the presence of suitable bases, there should be no loss of nitrogen whatsoever. If any loss of this element occurs, the process itself must be at fault. A careful nitrogen balance sheet has therefore been kept for a number of pits and heaps, which shows that under normal conditions no loss of nitrogen takes place (Table XX). On the contrary, nitrogen is gained, apparently by fixation from the atmosphere.

TABLE XX
NITROGEN BALANCE SHEETS IN NORMAL PITS AND HEAPS

No.	Description	Total N (lb.) at the beginning	Total N (lb.) in the finished product	Total gain in nitrogen	Percentage gain of nitrogen
Pit					
14	Standard (1/4 dung)	29.12	32.36	3.24	11.1
15	Full dung	32.70	34.87	2.17	6.6
16	Dry dung	30.41	32.33	1.92	6.3
18	Full dung (residues low in nitrogen)	29.10	36.77	7.67	26.3
19	Dry dung	29.55	30.70	1.15	3.9
20	Standard (1/4 dung)	24.73	25.80	1.07	4.3
21	Full dung (half period in monsoon)	32.35	33.40	0.15	0.45
42	Monsoon	22.28	29.52	1.24	4.4

THE WASTE PRODUCTS OF AGRICULTURE

In one case, No. 18, in which residues poor in nitrogen were composted with the full supply of dung, a very large amount of fixation took place. It will be interesting to investigate cases such as these in greater detail, and to determine the exact conditions under which such a large volume of free nitrogen can be fixed.

While losses of nitrogen do not take place in normal pits or heaps, water logging of the pits during the early rains, even for a short period, is at once followed by denitrification (Table XXI).

TABLE XXI
NITROGEN BALANCE SHEET OF TEMPORARILY
WATERLOGGED PITS

No.	Description	Total N (lb.) at the beginning	Total N (lb.) in the finished product	Total gain in nitrogen	Percentage gain of nitrogen
Pit					
24	Full dung	31.80	29.66	2.14	6.7
25	Full dung	29.55	27.10	2.15	8.1

Nitrogen is always lost in the first stage of the process--between charging and the first turn--whenever the nitrogen content of the mass is too high at the beginning (Table XXII).

TABLE XXII
CHANGES IN NITROGEN CONTENT DURING THE FIRST
STAGES OF THE PROCESS

No.	Description	Percentage Nitrogen at the beginning	Percentage Nitrogen after the first turn
	Residues poor in nitrogen		
Pit 14	Standard--dry season	0.68	0.84
Pit 25	Standard--dry season	0.63	0.60
Heap 41	Standard--monsoon	0.64	0.64
	Residues rich in nitrogen		
Pit 5	Full dung--dry season	1.04	0.70
Pit 6	Full dung--dry season	0.86	0.73
Pit 40	Full dung--monsoon	1.30	1.03

THE WASTE PRODUCTS OF AGRICULTURE

Another loss of nitrogen which has to be guarded against takes place when the final product is kept too long in heaps. An appreciable loss of nitrogen takes place even when the compost is kept for an extra month in the heap (Table XXIII). After ninety days the process is complete, when the humus should be used as a top dressing for growing crops or else banked by applying it to the land, when it becomes diluted with such large volumes of dry earth that all further changes are checked.

TABLE XXIII
NITROGEN LOSSES DURING STORAGE IN HEAPS

No. of Pit months	Percentage of total nitrogen on dry basis after three months	Percentage of total nitrogen on dry basis after four
7	0.90	0.88
8	1.00	0.93
14	0.84	0.81
15	0.72	0.68

THE CHARACTER OF THE FINAL PRODUCT

The ripe compost consists of a brownish-black, finely divided powder, of which about 80 per cent will pass through a sieve of six meshes to the linear inch. The state of division of an organic manure is an important factor, second only to its chemical composition. This property enables the Indore compost to be rapidly and easily incorporated, and to exert its maximum effect on the internal surface of the soil. The carbon-nitrogen ratio is not far from the ideal figure of 1O:1. The nitrogen is therefore in a stable form, which does not permit of liberation beyond the absorption capacity of the crop. The percentage of total nitrogen is also satisfactory, varying from 0.8 to 1.0 per cent (Table XXIV).

TABLE XXIV
COMPOSITION OF THE FINAL PRODUCT

No. of pit or heap	Materials used	Organic Matter	Total ash	Silicates & sand	Nitrogen	P2O5	K2O	C/N	Soluble humus	Fineness
Heap	Cotton-stalks with reduced (1/4) dung	33.92	66.09	34.97	1.61	0.48	3.38	16.5:1	11.56	68.15
Pit 7	Dry mixed residues	20.14	79.87	46.91	0.0	0.41	1.95	11.2:1	5.56	72.3
Pit 14	Dry mixed residues	19.66	80.34	46.32	0.84	0.68	2.35	11.6:1	6.27	88.5
Pit 8	Dry mixed with full dung	20.19	79.82	46.27	1.004	0.51	3.05	10.8:1	4.83	81.3
Pit 15	Dry mixed with full dung	18.39	81.62	51.33	0.725	--	2.43	12.6:1	3.86	82.5
Pit 5	Dry mixed with full dung	19.76	80.24	50.11	0.841	0.403	2.23	11.7:1	5.29	84.0

Results obtained in the monsoon

No. of pit or heap	Materials used	Organic Matter	Total ash	Silicates & sand	Nitrogen	P2O5	K2O	C/N	Soluble humus	Fineness
Heap 6	Mixed withered weeds	21.25	78.75	47.55	0.862	0.43	2.33	12.3:1	4.01	76.3
Heap 10	Mixed withered weeds	22.05	77.95	47.77	0.808	0.49	4.99	13.6:1	4.07	78.4
Heap 22	Mixed withered weeds	22.09	77.91	48.45	0.914	0.51	3.59	12:1	4.31	75.7
Heap 34	Mixed withered weeds	19.38	80.63	48.7	0.625	0.59	5.31	15.5:1	4.27	79.4
Heap 40	Half withered weeds, half sann	21.05	79.95	47.61	0.825	0.55	2.85	12.75:1	5.96	78.6
Heap 42	Dry mixed residues	21.69	78.32	46.41	0.806	0.62	3.65	13.5:1	5.36	84.0

The nitrifying power of the compost, particularly that made from mixed residues, is very satisfactory. Laboratory tests, carried out under conditions resembling those of the field during the early monsoon rains . . . bring out clearly the superiority of the product made from mixed residues.

Besides its value as a source of readily available nitrogen, the Indore compost acts as an indirect manure. The permeability of the black cotton soil is markedly improved, particularly by the product from mixed residues. The loss of permeability which takes place in these soils after the early rains, is perhaps the greatest obstacle to high yields of cotton. A manure, therefore, which will help to remove this factor, is exactly what the cultivator needs. This property will prove of the greatest value in keeping alkali in check, when the process is applied to the close alluvial soils of the Punjab and Sind.

THE WASTE PRODUCTS OF AGRICULTURE

It will be clear from the results set out in this chapter that a solution of the problem of utilizing the waste products of agriculture itself has been solved, by methods which are well within the means of any industrious cultivator. All the recent work on the problems of manuring points clearly to the supreme importance of organic matter of the right type. This must possess a carbon-nitrogen ratio in the neighbourhood of 10:1, and must be synthesized from crop residues by means of fungi and bacteria, working under aerobic conditions. Clearly the thing to do is to manufacture such a product in a compost factory under strict control, and then to add the organic matter to the soil. This has been accomplished at Indore.

NOTE: After this chapter was written, a paper by Waksman and Gerretsen appeared in the issue of Ecology of January 1931, which confirms the results set out above in a very remarkable way, The New Jersey experiments deal with the influence of temperature and moisture on the decomposition of plant residues as a whole. The higher the temperature, the more rapid is the decomposition of the material including the lignins. At the highest temperature, 37 degrees C., the carbon-nitrogen ratio was reduced from about 1OO to 11.3:1, to almost the ratio of the organic matter in normal soil. When decomposition was most favourable and most rapid, the final carbon-nitrogen ratio was practically the same as that in soil humus. This is exactly what happens in the Indore process. The American results, which were obtained under laboratory conditions, fully confirm our factory experience of the last four years in India and can be applied, practically as they stand, to the Indore process.

BIBLIOGRAPHY

BRAYNE, F. L.--The Remaking of Village India, Oxford University Press, 1929.

DUBOS, R. J.--'Influence of Environmental Conditions on the Activities of Cellulose Decomposing Organisms in the Soil,' Ecology, 9, 1928, p. 12.

HOWARD, A. and HOWARD, G. L. C.--The Application of Science to Crop Production, an Experiment carried out at the Institute of Plant Industry, Indore, Oxford University Press, 1929.

HUTCHINSON, H. B.--'The Influence of Plant Residues on Nitrogen Fixation and on losses of Nitrates in the Soil,' Journ. of Agric. Science, 9, 1918, p. 92.

JENSEN, H. L.--'The Microbiology of Farmyard Manure in Soil. 1--Changes in the Micro-flora and their Relation to Nitrification,' Journ. of Agric. Science, 21, 1931, p. 38.

RUSSELL, E. J. and RICHARDS, E. H.--'The Changes taking Place during the Storage of Farmyard Manure,' Journ. Of Agric. Science, 8, 1917, p. 95.

VILJOEN, J. A., FRED, E. B. and PETERSON, W. H.--'The Fermentation of Cellulose by Thermophilic Bacteria,' Journ. of Agric. Science, 16, 1926, p. 1.

VOELCKER, J. A.--Report on the Improvement of Indian Agriculture, London, 1893.

WAKSMAN, S. A--'The Influence of Micro-organisms upon the Carbon-Nitrogen Ratio in the Soil,' Journ. of Agric. Science, 14, 1924, p. 535.

WAKSMAN, S. A. and TENNEY, F. Q.--'Composition of Natural Organic Materials and their Decomposition in Soil. IV--The Nature and Rapidity of the various Organic Complexes in different Plant Materials under Aerobic Conditions,' Soil Science, 28, 1929, p. 55.

WAKSMAN, S. A. and DIEHM, R. A.--'Chemical and Microbiological Principles underlying the Transformation of Organic Matter in Stable Manure in the Soil,' Journ. of the American Soc. of Agronomy, 21, 1929, p. 795.

WAKSMAN, S. A. and GERRETSEN, F. C.--'Influence of Temperature and Moisture upon the Nature and Extent of Decomposition of Plant Residues by Micro-organisms,' Ecology, 12, 1931, p. 33.

WHITING, A. L. and SCHOONOVER, W. R.--The Comparative Rate of Decomposition of green and cured Clover Tops in Soil, Soil Science, 9, 1920, p.137.

VI. APPLICATION TO OTHER AREAS

In the present chapter, the various adaptations that will be needed, and the further investigations that must be undertaken before the Indore process can be widely adopted, will briefly be considered.

ADAPTATIONS

As far as the tropics and sub-tropics are concerned, the process can be adopted as it stands. No particular difficulties are likely to be encountered at any stage. After the collection, storage and admixture of the raw materials, including dung and urine earth, the two chief factors on which success depends are: (1) the maintenance of a high temperature in the pits or heaps; and (2) adequate aeration throughout the manufacture. With ordinary care, temperature difficulties are unlikely to occur, as the daily mean in these regions is always high, and the occasional cold spells are of short duration. All that is needed is the proper orientation of the pits or heaps to prevent overdue cooling by high winds, particularly during the interval between charging and the first turn. The maintenance of the correct degree of aeration requires more care. The chief difficulty likely to arise is the flooding of the pits after heavy rain or by the rise of the ground water. The material then becomes thoroughly soaked, and adequate aeration is impossible. If this over watering cannot be prevented by catch drains, pits will have to be given up and the manufacture conducted in heaps on the surface. Direct wetting through heavy falls does little or no permanent harm. This was clearly established at Indore during the monsoon of 1930, when the total rainfall was forty-five inches, most of which was received between I5 June and 15 September. This included five falls of over two inches and two of over five inches in twenty-four hours. In spite of these heavy downpours, the conversion proceeded evenly and without difficulty; there was little or no loss of soluble nitrogen by leaching; the amount of moisture absorbed from the rainfall did not interfere with the oxygen supply. For these reasons it is not necessary in warm countries to carry on the manufacture under cover. The erection and maintenance of sheds therefore need not be considered.

In the damper areas of the tropics like parts of Africa and the West Indies, which do not possess a cattle force at all comparable with that of India, a difficulty in maintaining the correct carbon-nitrogen ratio of the mixture may occur. There may be insufficient dung and urine earth for converting the large quantities of vegetable wastes which are available. The shortage can be made up by the use of nitrate of soda or by the Adco powders. If such artificials are employed, it will be a great advantage to make use of soil as the principal base for keeping the general reaction uniform and within the optimum range. Soil is the best base for neutralizing acidity and for absorbing ammonia and is far more effective than lime or wood ashes. This material possesses two other important advantages in the making of compost. In the first place, the soil colloids are very retentive of moisture and so help to keep the water content of the mass steady. In the second place, the colloids cover the vegetable matter with a thin adherent film which can retain in situ all the materials--combined nitrogen and minerals, soluble carbohydrates, water and oxygen essential for the rapid development of the micro-organisms. The result is that there is no delay in the breaking down of the vegetable wastes and in the synthesis of microbial tissue. When earth is omitted from the mixture, two difficulties at once arise. The supply of moisture for the microorganisms is intermittent; the general reaction becomes inconsistent. Delays ensue. For these reasons, the Adco process could easily be improved by the judicious use of earth. If lime were omitted from the Adco mixture, the freight on this item could be saved and the usefulness of the rest of the powder increased.

In those areas of the temperate regions where winter occurs, one important modification of the process may be needed. As will be evident from a study of the results set out in Chapter V, one of the difficulties against which provision has to be made is the lowering of the temperature of the fermenting mass by cold and wind. For the micro-organisms to complete the conversion in ninety days, the heaps must be kept at a high temperature throughout. No difficulties are likely to arise during the summer. Trouble however is likely during the colder months--November to April. During this period the fermentation may have to be carried out in sheds or in compost houses on the Japanese principle. Many existing farm buildings could be adapted for the pur-pose; the ideal structure however would have to be designed--a task which will be lightened after a careful study of the methods in use in those areas of Japan where compost houses are the rule.

The difficulty of adopting the system in countries like Canada, the United States and Great Britain, where labour is dear and scarce, will be solved by the mechanization of the process. The first step would be for one or two of the experiment stations to transform all their vegetable wastes into compost by hand labour regardless of ex-pense, and then to determine the value of the product in maintaining crop production at a high level. The full possibilities of humus will only appear when the dressings of compost are supplemented by the addition of suitable artificials. The combination of the two, applied at the right moment and in proper proportions, will open the door to the intensive crop production of the future. Humus and artificials will sup-plement one another. Further, the artificials must not be confined to those which merely supply nitrogen, phosphates and potash. Sub-stances like lime and sulphur, which flocculate the soil colloids and so improve the filth, must be included.

In other words, the manuring of the future wild have to be both direct and indirect.

FURTHER INVESTIGATIONS

In the tropics and sub-tropics, an important aspect of the proc-ess is its application to the future sanitation of the village. The fact that forty oxen are kept at the Institute of Plant Industry, Indore, and that compost is manufactured throughout the year, without the slightest smell and without the breeding of flies, indicates clearly the line of advance in dealing with village sanitation. All that appears to be needed is to adapt the Indore process (which employs cow-dung and urine earth) to the use of night soil, and to utilize the present sanitary services in showing the people how to transform the village wastes (including all forms of litter of vegetable origin) into compost. No difficulties are likely to be experienced in the actual conversion of the waste products of the rural population into humus. The process will be more rapid than when cow-dung is used: a factor which is all to the good. Besides the valuable compost that will be obtained, a number of other advantages will follow. Rural hygiene will enter on a new phase. The fly nuisance will disappear. Practically all the infection, which is now carried by these insects from filth to the food and water supply of the population, will be automatically destroyed by the combination of high temperature, high humidity and copious aeration of the compost heaps. In the tropics parasites like hookworm will tend to decrease in numbers. A rapid improvement in the general health and the amenities of the village will ensue. What is needed to bring about these results is the working out of a simple process on the lines of the one described in this book. It will not prove a difficult. It will be easy to-design a series of screened pits and screened areas in the neighbourhood of an Indian village, and to teach the sweepers how to carry on the manufacture of compost without smell and without the breeding of flies. The conditions which render these two nuisances impossible will at the same time destroy practically all the harmful parasites and germs which now infect the population. Provided the work is carried out by the village scavengers, no caste difficulties are likely to arise. The process can easily be welded into the existing village system. A beginning has been made in the direction indicated by Mr. F. L. Brayne, I.C.S., Deputy Commissioner of Jhelum (formerly Deputy Commissioner of Gurgaon). Mr. Brayne has designed a latrine pit, which without much difficulty could be perfected for use throughout the tropics and sub-tropics. The method will have to be adapted both to dry weather and to monsoon conditions, and will have to be worked for a year or two under strict microbiological and chemical control before being brought to the notice of the people. Work on these lines has already been started in the model village belonging to the Institute of Plant Industry at Indore. If, as seems certain, a practicable method can be devised, steps will at once be taken to get it taken up in the villages of the Central India and Rajputana States. Its spread to the rest of India, and all over the tropics and subtropics, will be a matter of a very few years.

The moment a suitable method of dealing with the sanitation of the village has been designed and the influence of the process on the general health of the people and on the fertility of the fields becomes manifest, the results can be carried further. The public health of the military cantonments and of the smaller towns can then be considered as one subject. In place of the present expensive division of those aspects of the general problem of sanitation, which deal with solid wastes, into a number of imperfectly related items, such as--the disposal of night soil, the use of disinfectants, the collection and destruction by burning of vegetable wastes including fallen leaves, the prevention of the fly nuisance, the purification and safeguarding of the water supply and the inoculation of the population against such diseases as enteric fever and cholera--it will be possible to transform these waste products of the population into valuable humus in a scientific way, and so avoid most if not all the existing difficulties. Such results, as far as urban areas are concerned, will naturally be the work of years. In the villages, however, progress should be rapid. The first important step on the road has already been taken in the form of the Indore process. It will not be a difficult matter to expand the opening which has been made. Little increase in public expenditure will be called for. The funds and staff, now devoted to rural hygiene, can at once be deflected to the manufacture of compost and to increasing the produce of the soil.

BIBLIOGRAPHY

BRAYNE, F. L.--The Remaking of Village India, Oxford University Press, 1929.

HALL, A. D.--' Some Secondary Actions of Manures upon the Soil,' Journ. of the Royal Agric. Soc. of England, 70, 1909, p. 12.

HOWARD, A. and HOWARD, G. L. C.--The Application of Science to Crop Production, an Experiment carried out at the Institute of Plant Industry, Indore, Oxford University Press, Bombay, 1929.

KING, F. H.--Farmers of Forty Centuries or Permanent Agriculture in China, Korea and Japan, London, 1926.

Appendices

A. THE MANURIAL PROBLEM IN INDIA

The manurial problems of India were considered in detail by the recent Royal Commission on Agriculture in India which, after an extensive study of the subject lasting more than two years, reported in 1928. That section of the report which deals with fertilizers is reprinted in full below. A study of this account will bring home to the investigator and to the general reader the evils which invariably result from the fragmentation of any large agricultural problem.

(Extract from the Report of the Royal Commission on Agriculture in India, Bombay, 1928, pp. 80-93.)

FERTILIZERS

80. 'Of the principal plant-food materials in which the soils of India are deficient by far the most important (except in parts of the crystalline tracts where the deficiency of phosphates may be more serious) is nitrogen, and the manurial problem in India is, in the main, one of nitrogen deficiency. India, as is well known, depends almost exclusively on the recuperative effects of natural processes in the soil to restore the combined nitrogen annually removed in the crops, for but little of this is returned to the soil in any other way. Much of the farmyard manure available is burnt as fuel whilst a large quantity of combined nitrogen is exported in the form of oil seeds, food and other grains, and animal products such as hides and bones. This loss is in no way compensated by the importation of nitrogenous fertilizers, for 1925-26 was the first year which the imports of sulphate of ammonia into this country, which amounted only to 4,724 tons, exceeded the

exports and was also the first year in which the greater part of the production of this fertilizer by the Tata Iron and Steel Company at Jamshedpur and in the coalfields of Bengal and Bihar and Orissa was consumed in India. In these circumstances, it is fortunate that the recuperative processes in the soil are more pronounced in tropical and subtropical than in temperate regions. Although it has been stated in evidence before us that it has not been established that improved and higher yielding varieties of crops, more especially of wheat and sugarcane, take more from the soil than the varieties they replace, and that their cultivation on present lines will not, therefore, be followed by any loss of permanent fertility, we are of opinion that there is justification for the view that improved crops generally require, for their fullest development, more liberal manurial treatment than those ordinarily grown. The subject is one which requires careful study by the agricultural departments in India and should form an essential part of the investigations discussed in the following paragraph.

MANURIAL EXPERIMENTS

81. An acceleration of the recuperative processes in the soil can be effected by improved agricultural methods, by adequate soil aeration, judicious rotations and the cultivation of green-manure crops. The loss of combined nitrogen can also be partially made up by the application of natural and artificial manures. With certain definite exceptions, however, such as, for instance, sugar-cane and the more valuable garden crops, it has yet to be determined for what conditions and for what crops artificial manures can be profitably used to stimulate crop production in India. In this connexion, we have been impressed by the importance of research into the fundamental problems connected with losses in nitrogen and with nitrogen recuperation. We saw something of the work in this field which was being carried on at Pusa by Dr. Harrison and at Nagpur by Dr. Annett. Although, ever since the reorganization of the agricultural departments in 1905, manurial experiments have engaged a large part of their time and energies and have been carried out on every agricultural station in India, it cannot be said that the agricultural experts are even yet in a position to give satisfactory advice to the cultivator in regard to the use of manures. A large amount of data has been collected but it has not been studied systematically or reduced to a form which would enable clear and definite conclusions to be drawn. The problem requires to be stud-

ied in three aspects: in relation, in the first instance, to the crops which are dependent solely on rainfall, in the second, to crops which are grown on irrigated land, and lastly, to the planters' crops and intensive cultivation such as that of sugar-cane and garden crops. It is hardly necessary to point out that the use of nitrogenous or other artificial fertilizers is not profitable in all conditions. Where crop production is limited by a small rainfall, the annual additions of combined nitrogen to the soil as the result of natural processes may be sufficient to meet the needs of a crop the out-turn of which is limited by the moisture available. It has, for example, been found in the Central Provinces that the application of fertilizers benefits dry crops, including unirrigated cotton, only in years when the rainfall is adequate and that, in particular, it does not benefit wheat which, in that province, is grown on rainfall only. The planting community, which has its own specialist officers, needs no advice from the agricultural departments in regard to the economic use of manures. We would, however, take this opportunity of stressing the value of close touch between the community and the departments in regard to this and other agricultural matters It is essential that the departments should be in a position to give the ordinary cultivator, both of irrigated and unirrigated crops, definite guidance on the point. The first step is the careful study of the existing material and the correlation of the results hitherto obtained. The second step is the formulation of a programme of experiment with the object of ascertaining, with all possible accuracy, the extent to which fertilizers can be used with profit. This programme should include the laying out of a short series of permanent manurial plots, on lines appropriate to conditions in India, on provincial experimental farms. Only by conducting manurial experiments over a number of years will it be possible to compile such records as would make a substantial contribution to the knowledge of the problems of manures and manuring under tropical and sub-tropical conditions about which little is yet known. The scientific value of continuous experiments depends on accurate methods of collection of all relevant data with a view to their subsequent correlation. All such schemes for manurial trials would ordinarily be drawn up by the Director of Agriculture in close consultation with the agricultural chemist and the deputy directors of agriculture under whose immediate supervision the experiments would be conducted. We wish especially to emphasize the importance of manurial experiments on unirrigated land as the cultivator of such land, who runs, with his very limited financial resources, the risk of losing his crop in an unfavourable season, stands most in need of guidance in this matter. The study of the available data and the formulation of an ordered programme to replace the present somewhat haphazard methods of dealing with the problem would, we think, provide sufficient work to justify an officer of the Agricultural Department being placed on special duty for a limited period, but we prefer to make no definite recommendations on this point and to leave it to the consideration of the local governments. Local conditions vary so greatly between province and province, especially in regard to unirrigated land, that it does not appear necessary to attach an officer to Pusa specially to assist the provinces in this investigation. The Council of Agricultural Research should be in a position to advise as to the manner in which the experiments can best be conducted so as to secure uniformity of method as far as possible and to render the results obtained in one province of some value to other provinces.

INTERNAL SOURCES OF SUPPLY AND THEIR DEVELOPMENT

(a) FARMYARD MANURE

82. The first question which arises, in considering the internal supplies of nitrogen available in India and the methods by which these can best be developed, is that of the use of farmyard manure as fuel. The view is generally held that it is the absence of a sufficient supply of firewood which, over large parts of India, compels the burning of cow-dung as fuel. But it must be recognized that there is often a definite preference for this form of fuel, as its slow burning character is regarded as making it specially suitable to the needs of the Indian housewife. Thus we are informed that, in Burma, immigrant labourers from India persist in using cow-dung as fuel although an abundant supply of firewood is readily available. Our evidence does not suggest any alternative fuel for domestic purposes in districts where wood and coal are dear. In some tracts, cotton-stalks, the dry stubble and stalks of tur (Cajanus indicus), the pith of jute and sann hemp and the bagass of sugar-cane, where the use of the McGlashan furnace leaves a surplus which is not required for boiling the juice, could be utilized for fuel to a far greater extent than they are at present. Fuel plantations, more especially irrigated plantations, the formation of which we discuss in Chapters VIII and X, can assist only in a very limited area. In our view, the agricultural departments have a difficult task to perform

in attempting to promote the utilization of farmyard manure for its proper purpose. Propaganda in this direction can only prove effective if an alternative fuel is suggested and if the cultivator can be sufficiently imbued with a sense of thrift to induce him to burn that which will probably seem to him a less satisfactory substance. There has been little advance in regard to the preservation of manure since Dr. Voelcker wrote his report on Indian agriculture in 1893. The practice of providing litter for cattle is rarely, if ever, adopted except on government farms. No efforts are made by the cultivator to preserve cattle urine. Manure pits are still seldom found in Indian villages. Where they do exist, no attempts are made to preserve the manurial value of the contents or to safeguard the public health by covering the material with earth.

(b) COMPOSTS

83. While the task is difficult, there is no doubt that something can be done to promote the better preservation of such farmyard manure as is not diverted to consumption as fuel, by using it as a compost with village sweepings, leaves, and other decomposed vegetable matter. In this connexion, we are impressed by the results achieved in the Gurgaon district of the Punjab, where many villages have, as a direct consequence of propaganda, adopted the practice of depositing in pits all village sweepings and refuse, along with a proportion of cow dung. The effects on crops to which such manure has been applied, and on the sanitation and general amenities of the villages, were most marked. There is no reason why efforts on similar lines should not be made in other parts of the country. The Indian cultivator has much to learn from the Chinese and the Japanese cultivator in regard to the manufacture of composts. Artificial fertilizers are used as little in China as they are in India; but there is no organic refuse of any kind in that country which does not find its way back to the fields as a fertilizer. Not only is all human waste carefully collected and utilized, but enormous quantities of compost are manufactured from the waste of cattle, horses, swine and poultry, combined with herbage straw, and other similar waste. Garbage and sewage are both used as manure. The agricultural departments in India are fully alive to the necessity for instructing the cultivator in the better preservation of manure and the use of composts, but there is great scope for an extension of their activities in this respect. For example, the possibilities of manufacturing

synthetic farmyard manure from waste organic material on the lines worked out at Rothamsted deserve to be fully investigated. At Rothamsted, research was at first directed towards discovering artificial means whereby the decomposition of straw might be effected. Straw contains three essentials to plant growth, viz. nitrogen, phosphate and potash. The work proved successful and a method was devised for treating large quantities of straw for the preparation of manure. Reagents were subsequently discovered which were capable of bringing about the rapid rotting, not only of straw but also of other plant residues, and thus of producing a valuable organic manure at a moderate cost. Synthetic farmyard manure is being prepared by the departments of agriculture in Madras and the Central Provinces. The agricultural department in Bengal, following the valuable lead given by Rothamsted, has attempted the manufacture of artificial farmyard manure on a considerable scale. Cattle urine and washings from cattle-sheds, mixed with bone meal, have been used with immediate success. Weeds, various grasses, sugar-cane trash, refuse, straw, prickly-pear, etc., have all proved capable of breaking down into excellent material approximating more or less closely in appearance and in composition to that of cow-dung. Experiments have also been made in Burma but have not so far proved successful. Valuable work on the preparation of composts from night soil and refuse and from cattle urine, weeds, etc., is being done by Dr. Fowler at Cawnpore. In Europe, work of this character has now emerged from the experimental stage and processes devised for dealing with various classes of materials are already on the market. In India, however, the departments concerned have still to devise and introduce a practical method which can be used with profit by the ordinary cultivator on his own land.

The manurial value of earth obtained from the sites of abandoned villages is recognized in many parts of India. The quantities available are, however, negligible in relation to the manurial requirements of the country.

(c) NIGHT SOIL:

84. Prejudice against the use of night soil has deterred the cultivator in India from utilizing to the best advantage a valuable source of combined nitrogen. There is, however, evidence that this prejudice is weakening and that, where night soil is available in the form of

poudrette, it is tending to disappear. From the point of view of public health, the use of poudrette is preferable to that of crude night soil and, given co-operation between agricultural departments and municipal authorities, there is hope that the manufacture of poudrette should prove profitable to municipalities and beneficial to the cultivators in their neighbourhood. The methods of converting night soil into poudrette adopted at Nasik and elsewhere in the Bombay Presidency have been highly successful and appear well worth study by other municipalities. The advantages of this system of dealing with night soil appear to us to justify a recommendation that the departments of local self-government in all provinces should bring them to the notice of all municipal authorities and should also take steps to establish a centre at which members of the municipal sanitary staffs can receive a suitable training in this method of disposing of night soil. The agricultural departments should keep a watchful eye on all experiments in the conversion of night soil into manure and should themselves conduct such experiments. Where municipal authorities in any part of the country are in a position to supply it, the agricultural departments should assist them to find a market by arranging demonstrations of the value of night soil as manure on plots in the neighbourhood of the towns.

Another way in which night soil can be converted into a form in which its use is less obnoxious to the cultivator is by the adoption of the activated sludge process. This process reduces sewage, by the passage of air through it, to a product which can either be used as required in the form of effluent from the sewage tanks or dried and sent where there is a demand for it. The activated sludge process is suitable only for towns which have a sewage system. It is much more expensive than conversion into poudrette but has the advantage of conserving a larger percentage of nitrogen. Up to the present, this system has been adopted in India on any considerable scale only at Tatanagar. The possibility of selling the product at a price that would yield a fair return on the cost of manufacture must depend upon a careful survey of all the relevant factors, including the local market for the product. In estimating the cost of the necessary plant, due regard should be paid to the cost which would be involved in installing any alternative method of sewage disposal, and, if it should prove possible to place a valuable fertilizer at the disposal of the cultivators at a price they can afford to pay, without risk of imposing any additional net charge upon the local ratepayers, we think that it is in the public interest that such schemes should be adopted.

(d) LEGUMINOUS CROPS

85. Another indigenous source of combined nitrogen to which increasing attention is now being paid by the agricultural departments in India, is leguminous crops and green-manures. The value of leguminous crops in his rotation has always been recognized by the cultivator and the work before the agricultural departments in regard to these crops lies not so much in popularising the principle of their cultivation as in discovering the varieties of leguminous crops best suited to increase the soil fertility and in recommending such varieties to the cultivators. Recent research has drawn attention to the fact that such crops vary greatly in their power of fixing nitrogen in the soil and should not be regarded as of equal value. Moreover, it is only when the leguminous crop is grown for green-manure that, in all cases, the soil gains in nitrogen. Mr. Howard instances gram as a crop which improves the soil and Java indigo as a crop which seriously depletes the supply of combined nitrogen.

(e) GREEN-MANURES

86. The agricultural departments in India have devoted much time and attention to work on green-manure crops with a view to discovering the crops which can best be used for green-manure, the time at which they should be grown and the manner in which they should be applied. Their work has shown that sann hemp on the whole gives the best result and it would doubtless be more often grown for use as green-manure were it not that it may exhaust so much of the moisture in the soil that, when it is ploughed in, there is not sufficient left both to decompose it and to enable a second crop to grow. Much experimental work is still, therefore, required to discover the green-manure crops which can best be included in the cultivators' rotations. The economics of green-manure crops from the point of view of the small cultivator also require to be worked out. The small cultivator is naturally hesitant about growing a crop which only indirectly brings him any financial advantage. With his slender resources, it is indeed not unreasonable for him to take the view that he cannot afford to sacrifice even a catch crop in this way and it is therefore not until the agricultural departments are in a position to demonstrate to him beyond a

shadow of doubt the paying nature of green-manure crops on small holdings that these departments will be justified in persuading the small cultivator to adopt them or that their advocacy of them will stand any chance of success. In the present state of knowledge, such crops would appear an expedient to be adopted by the larger landholder and, for the small cultivator, a leguminous crop in his rotation would seem to hold out better prospects of benefit.

The possibility of growing such crops as dhaincha and ground-nut, the leaves of which can be used as green-manure without interfering with the commercial value of the crop, is worth consideration. The use of ground-nut in this way for green-manure would furnish an additional reason for extending the area of this valuable crop. In the case of crops of a woody nature such as sann hemp, it must, however, be remembered that their utility as green-manure for the succeeding rabi crops depends to a large extent on the presence of sufficient moisture in the soil to rot the dry stems and roots.

In Madras, the Punjab and the Central Provinces, the experiment has been made of encouraging the cultivation of green-manure crops under irrigation by the remission of the charge for water from government sources or irrigation. The fact that the results have so far been disappointing may be due to a failure to accompany the remission with sufficient propaganda as to the advantages to be derived from the growing of these crops. We think that the continuance of the concession and its extension to other areas should be conditional on its being accompanied by an active campaign of propaganda, directed particularly to the larger landholder rather than the small cultivator. All areas where the concession is made should be kept under regular examination. If, after a period of five to ten years, it should appear that the concession given in regard to water charges has failed to achieve its main purpose, it should be rescinded.

(f) 0IL CAKES

87. The loss to India of a valuable source of combined nitrogen as the result of the export of so large a proportion of its production of oil seeds was emphasized by many witnesses before us. The [figures for] yield and exports of oil seeds during the last fifteen years . . . . indicate that, of the out-turn of the seed of cotton, ground-nut, rape

and mustard, linseed and sesamum, the exports amount to an average of eighteen per cent and they suggest the loss which the soil of India suffers by the export of a valuable by-product on the assumption that the whole of the nitrogen contained might be returned to the soil. Under existing practice, indeed, much of this material would probably be fed to cattle and subsequently dissipated as fuel. But it is not surprising that the view that an export tax on oil seeds and oil cakes within the purchasing power of the cultivator has found much favour and even received the support of the Board of Agriculture in 1919 and of the majority of the Indian Taxation Enquiry Committee, but not that of the Indian Fiscal Commission. Some witnesses before us went further and urged the total prohibition of export. Whilst we fully recognize the advantages to Indian agriculture which would follow from a greatly extended use of certain oil cakes as a manure for the more valuable crops such as sugar-cane, tobacco, cotton and tea, we cannot but feel that those who suggest the attainment of this object by the restriction or prohibition of exports have failed to realize the economic implications of their proposal. In the first place, it must be remembered that India has no monopoly of the world's supplies of oil-seeds and is not even the chief supplier of those seeds. The world's linseed market is controlled by the Argentine crop and the sesamum market by the Chinese crop. The competition of West Africa in the supply of edible oils is becoming increasingly serious. In these circumstances, it is an economic axiom that an export duty will be borne by the producer and that the cultivator will, therefore, receive a lower price for the oil seeds exported. The acreage under oil seeds in British India is still considerably below the pre-war level and the tendency to replace oil seeds by other crops which may be inferred from this would undoubtedly be greatly accentuated if any effective restrictions on export were imposed. The immediate fall in price, which would result from such restrictions, would tend to a reduction of area and consequently of outturn. Even if such a fall in prices were obtained by the method advocated, the gain to the cultivator qua consumer would be far more than counterbalanced by the disadvantage to the cultivator qua grower by the loss of the income he at present derives from his export market. In the second place, it may be argued that if the Indian oil-crushing industry were fully developed to deal with the present out-turn of oil seeds, then the area might remain at its present level and there would grow up a considerable export of oil, while the cake would remain to be used as a feeding stuff or manure. The market for oil in this country is, however, a very limited one and will remain so until India has reached a more advanced stage of industrial development. The oil-crushing industry would, therefore, have to depend mainly on the export market for the sale of its main product. The problem of cheap and efficient transport to the great industrial centres of the west presents almost insurmountable difficulties. Oil-crushers in India would find themselves in competition with a well-established and highly efficient industry and there is little reason to believe that their costs of production or the quality of their product would enable them to compete successfully with that industry. In the third place, even if restriction on exports succeeded in reducing the price of oil cakes, this would mean that a section of the agricultural community would be penalized for the benefit of another and much smaller section, for the growers of oil seeds would probably not be those who would make the most use of the oil cakes.

A similar line of reasoning applies to oil cakes, the average exports of which from India for the five years ending 1925-26 were 165,600 tons, against a negligible import. The oil cakes exported from India are a far less important factor in the world's supply than are the oil seeds and, in these circumstances, the burden of the duty would be entirely borne by the producer, in this case the crushing industry. There can, in our view, be little doubt that the effect of a duty on oil cakes, with or without a duty on oil seeds, would be the curtailment of oil-crushing activities and a diminution in the available supply of oil cakes, in other words, it would have effects entirely different from those desired by its advocates. It is not, therefore, by any restriction on trade that Indian agriculture is likely to reap greater advantages from the supply of combined nitrogen available in the large crops of oil seeds she produces. The only methods by which these advantages can be secured are by the natural development of the oil-crushing industry coupled with great changes in cattle management and in the use of fuel. The question how far the development of the industry can be promoted by Government assistance in the matter of overcoming difficulties of transport and in the form of technological advice in regard to improved methods of manufacture and standardization is one for the departments of industries rather than the departments of agriculture. An extension of the oil-crushing industry would undoubtedly tend to promote the welfare of Indian agriculture and we would commend the investigation of its possibilities to the earnest consideration of all local governments.

(g) SULPHATE OF AMMONIA

88. The important potential sources of supply of combined nitrogen discussed in the preceding paragraphs are supplemented to a small though increasing extent by the sulphate of ammonia recovered as a by-product from coal at the Tata Iron and Steel Company's works at Jamshedpur and on the coalfields of Bengal and Bihar and Orissa. There has been a very marked increase both in the consumption and production of this fertilizer in India in recent years. Of the 4,436 tons produced in 1919, all but 472 tons were exported and there were no imports. In 1925, of the estimated production of 14,771 tons, 6,395 tons were retained in India. With three exceptions, all the producers of sulphate of ammonia in India have joined the British Sulphate of Ammonia Federation which, through its Indian agents, is conducting active propaganda to promote the use of artificial fertilizers and has established a number of local agencies in agricultural areas in several provinces. The manner in which this source of supply is being developed is very satisfactory and it is still more satisfactory that a market for increasing quantities of the sulphate of ammonia produced in India is being found in the country. The importance of the price factor need hardly be stressed, for though the present average price of Rs. 140 per ton free on rail at Calcutta is much lower than that which prevailed immediately after the War, it is sufficiently high to preclude the application of sulphate of ammonia to any except the most valuable of the cultivators' crops, such as sugar-cane or garden crops.

(h) ARTIFICIAL NITROGENOUS FERTILIZERS

89. A method of increasing the internal supplies of combined nitrogen in India, the adoption of which has received powerful support, is the establishment of synthetic processes for obtaining combined nitrogen from the air in forms suitable for use as fertilizers. The Indian Sugar Committee was of opinion that, from the point of view of the development of the sugar industry alone, the successful introduction of synthetic processes in India was a matter of the first importance. That Committee recommended that the possibilities of utilizing the hydro-electric schemes, which were at that time under investigation in the Punjab and the United Provinces, for the fixation of nitrogen should be thoroughly examined and that, if it were found

that electric energy could be obtained at a rate approximating to Rs. 60 per kilowatt year, a unit plant of sufficient size to afford trustworthy information should be installed. Of the three processes in use for the fixation of atmospheric nitrogen, the arc process, the cyanamide process and the manufacture of ammonia by direct synthesis, the Committee considered the cyanamide process as the one which offered the best prospects of success in India but drew attention to the possibilities of the Haber process for obtaining synthetic sulphate of ammonia.

The position has changed greatly since the report of the Sugar Committee was written. The full effects of the diversion of the capital, enterprise and, above all, the research devoted to the manufacture of munitions to the production of peace time requirements, had not been felt in 1920. Since then, it has resulted in a fall in the world's price of nitrogen by fifty per cent, and there are prospects of still lower prices in the near future. We see no reason to question the view which was placed before us in the course of the evidence we took in London that, in present circumstances, only very large units with a minimum capacity of about 150,000 tons of pure nitrogen per annum can be expected to pay even under the most favourable conditions in Great Britain and on the Continent of Europe and that conditions in India make it much less likely that even a unit of that capacity would prove a paying proposition. The possibilities of manufacturing nitrogen from the air in India have already been exhaustively examined by a leading firm of chemical manufacturers in England, which has decided against proceeding with the project. It is probable that no factory on a scale which could be contemplated by any local government, or even by the Imperial Government, would be in a position to produce synthetic nitrogenous fertilizers at a price less than that at which they can be imported. The whole object of establishing such a factory, that of producing fertilizers at a price which would place them within the reach of a far greater proportion of the agricultural community than is at present in a position to use them, would be defeated if a protective duty were imposed to enable its out-turn to compete against imported supplies. It is also to be hoped that, should the demand for artificial fertilizers in India make it worth while, private enterprise will come forward to erect synthetic nitrogen works in this country. While the economics of the industry remain as they stand to-day, we are unable to recommend any further investigation into the subject under government auspices.

CENTRAL ORGANIZATION FOR RESEARCH ON FERTILIZERS

90. The discussion of the question of nitrogenous fertilizers would not be complete without mention of the proposal placed before us by the British Sulphate of Ammonia Federation, Ltd., and Nitram, Ltd., for the establishment by the Government of India of a central fertilizer organization on which the Imperial and provincial agricultural departments as well as the important fertilizer interests would be represented. The two companies, which are already spending 23,000 pounds annually on research and propaganda in India, expressed their willingness to increase this amount to 50,000 pounds, the additional amount to be handed over to a central organization constituted in the manner they suggest, provided that an equal sum is contributed by Government. The companies have made it clear that the research and propaganda they contemplate would be on the use of fertilizers generally and would not in any way be confined to that of the products they manufacture or sell. This offer, though not disinterested is undoubtedly generous and we have given it our most careful consideration. We regret, however, that we are unable to see our way to recommend its acceptance. We cannot but feel that, whatever safeguards were imposed, the work of, and the advice given by, an organization, at least half the cost of which was borne by firms closely interested in the subject matter of the investigation, would be suspect and would thus be deprived of much of its usefulness, especially since, as we have pointed out, the agricultural departments in India are not yet in a position to pronounce authoritatively on the relative advantages of natural and artificial fertilizers. We, therefore, consider it preferable that the agricultural departments should remain entirely independent in this matter but we need hardly say that we would welcome the establishment by the two firms mentioned, or by any other fertilizer firms, of their own research stations in India working in the fullest co-operation with the agricultural departments, the Indian Tea Association, the Indian Central Cotton Committee and any other bodies interested in the fertilizer question. So much work remains to be done on the manurial problems of India that it is desirable that every possible agency should be employed on it. To the supply by the fertilizer interests of free samples for trial by the agricultural departments there can, of course, be no objection, but we do not consider that any financial assistance beyond what is involved in this should be accepted. In coming to this conclu-

sion, we have not overlooked the fact that the Rothamsted Experimental Station accepts grants from fertilizer interests to meet the cost of experiments with their products. Rothamsted is not, however, a government institution and, further, the experiments it carries out are only undertaken on the clear understanding that the information obtained is not to be used for purposes of propaganda. The conditions at Rothamsted are thus entirely different from those under which it is proposed that the central fertilizer organization in India should function.

BONES AND BONE MEAL

91. Nitrogen deficiency can be remedied to some extent by the application of bones and bone meal. This form of fertilizer is, however, of greater value as a means of rectifying the deficiency of phosphates which, as we have pointed out, is more prominent in peninsular India and Lower Burma than that of nitrogen. As with other forms of combined nitrogen, an important quantity of this fertilizer is lost to India by a failure to apply it to the soil and by export. Except in the War period, the total export of bones from India has shown little variation in the last twenty years. The average exports for the five years ending 1914-15 were 90,452 tons, valued at RS.64.20 lakhs. For the five years ending 1924-25 they were 87,881 tons, valued at Rs. 95.94 lakhs. In 1925-26 they were 84,297 tons valued at Rs. 89.16 lakhs and in 1926-27 100,005 tons valued at Rs. 97.76 lakhs. The imports of bone manures are negligible. Practically the whole of the exports are in the form of the manufactured product, that is in the form of crushed bones or of bone meal, the highest figure for the export of uncrushed bones in recent years being 545 tons in 1924-25. Only a very small proportion of the bone manure manufactured in India is consumed in the country. During the War period, when prices were low, freight space difficult to obtain and export demand weak, it was estimated that not more than ten per cent of the total production was consumed in India, and this at a time when the prices of all Indian agricultural produce were exceptionally high. Enquiries we have made show that there is no reason to believe that the percentage retained for internal consumption has increased since the close of the War. Many witnesses before us advocated that the heavy drain of phosphates involved in the large export of bones from this country should be ended by the total prohibition of exports and this proposal received the support of the Board of Agriculture in 1919, whilst the majority of the

Indian Taxation Enquiry Committee recommended the imposition of an export duty. For much the same reasons as those for which we have rejected the proposal for an export duty on oil seeds and oil cakes, we are unable to support this recommendation. As was pointed out by the Board of Agriculture in 1922, local consumption, even in the most favourable conditions in recent years, has accounted for such a small fraction of the total production that the industry could not continue to exist on that fraction, and the imposition of an export duty would involve a serious danger of its extinction through the closing down of its markets. Further, any restrictions on export would deprive one of the poorest sections of the population of a source of income of which it stands badly in need.

For slow growing crops such as fruit trees the rough crushing of bones is sufficient, but for other crops fine grinding is required. The crushing mills are at present located almost entirely at the ports and, in order to get bone manures to the cultivator, the establishment of small bone-crushing factories at up-country centres where sufficient supplies of bones are available has been advocated. A far more thorough investigation of the economics of the bone-crushing industry than has yet been carried out is, we consider, required before the establishment of such mills can safely be undertaken by private enterprise. The first essential is to obtain definite data in regard to the price at which, and the crops for which, the use of bone meal is advantageous to the cultivator. We suggest that the agricultural departments should take early steps to collect these data. The department of Government responsible should also investigate the cost of processing bones with special reference to those districts in which the development of hydro-electric schemes gives promise of a supply of cheap power. It should then be a comparatively easy matter to determine whether the level of prices is such as to justify any attempts on the part of Government to interest private, or preferably co-operative, enterprise in the establishment of bone-crushing mills in suitable centres. In determining the level of prices, allowance should be made for the advantage which local mills will enjoy in competition for local custom with the large units at the ports through the saving to the local concerns of the two-way transportation charges borne by the product of the port mills.

THE WASTE PRODUCTS OF AGRICULTURE

FISH MANURES

92. Little need be said about fish manures which are another source of supply of both phosphates and nitrogen. The export of these from India for the five years ending 1925-26 averaged 16,774 tons valued at Rs. 19.94 lakhs. In 1926-Z7 only 7,404 tons were exported valued at Rs. 9.21 lakhs. Except for a negligible export from Bombay and Sind, the exports of fish manures are confined to the west coast of Madras and parts of Burma.

The arguments against the prohibition of the export of bones or for the imposition of an export duty apply equally to fish manures. Any restriction of export would involve most serious hardship on the small and impoverished fishing communities of the two provinces, and cannot, therefore, be justified. The only measures which can be undertaken to lessen the export of fish manures, without damage to the fish-oil industry or the curtailment of the amount of fish caught, are measures to establish that such manures can be profitably used for Indian agriculture at the price obtained for them in the export market.

NATURAL

93. Reference should be made here to the extensive deposits of natural phosphates which are to be found in the Trichinopoly district of Madras and in South Bihar. In neither tract do these phosphates exist in a form in which they can be utilized economically for the manufacture of super phosphate; and their employment in agriculture has been limited to applications of the crude material in pulverized form. This source of supply does not offer any important possibilities.'

B. SOME ASPECTS OF SOIL IMPROVEMENT IN RELATION TO CROP PRODUCTION

By G. CLARKE, C.I.E., F.I.C., M.L.C. (Proceedings of the Seventeenth Indian Science Congress, Asiatic Society of Bengal, Calcutta, 1930, p. 23.)

I ASK your permission to direct your attention to some aspects of soil improvement in relation to crop production. I propose to pass in brief review some of our problems and then to touch on the work to which my colleagues, Khan Sahib Sheikh Mohammad Naib Husain, Rai Sahib S. C. Banerjee, and myself have devoted a number of years at the Shahjahanpur Research Station. My subject is directly connected with the supply of the first necessity of life, namely, food. By what methods is the world going to continue to feed its growing population? It is increasing at the rate of nearly twenty millions a year, and it cannot be suddenly checked. Can food be found for all these extra mouths, or will the pressure on our land resources become unbearable, and end in disaster? That is the colossal problem facing the world in the next few generations. It must be met either by a continual expansion of cultivation, or an intensification of production on land already cultivated.

How do we stand in India in respect to these questions? I have proceeded in a somewhat empirical fashion to ascertain the relation between population and arable land. I have selected, in making my estimate, the figures used in international statistics, the total area sown and the current fallows. I have deducted the area required for the production of exported cotton, food grains, oil seeds, jute, and tea, which account for about eighty per cent of the value of our exports. This estimate is admittedly rough and must be regarded as suggestive rather than as an exact measure, but it is sufficiently near to illustrate my point. I have taken the year 1922-23, following the census year 1921, and the year 1925-26. In 1922-23 the total area sown in that part of India for which agricultural returns are made was 327 million acres, 61 were under fallow, making a total of 388 million acres. From this may be deducted as producing exported material for cotton 14, for food grains 9, for oil seeds 5, for jute 2 for tea 0.6 million acres, or 31 million acres in round numbers. So that 357 million acres are left to supply the requirements in home produced food and other essential commodities of the 292 million people who live in the territory covered by these figures, viz. 1.2 acres per unit population.

A similar calculation for 1925-26 gives the same result. I have selected for a summary comparison the United States of America and France, two countries possessing points of resemblance to India. In both, as in India, agriculture is of predominant importance. In the United States 356 million acres are in cultivation: 65 million producing exported material may be deducted from this, leaving 291 million acres of cultivated land devoted to supplying a population of approximately 112 million, or 2.6 acres per unit of population. The dominant characteristic of American economic life has hitherto been abundance of land resources. France, a country which is largely self-supporting, has 36.3 million hectares of cultivated land for a population of 39.3 million, approximately 2.3 acres for each head of the population.

In considering these figures we have to allow for the fact that the vegetarian diet adopted by our people is more economical of the resources of the soil than the diet of the people of the United States and France. Living is cheap in India, but when all has been said that can be said, we are left with the plain fact before us that we have one-half the area of cultivated land for a unit of population.

The past experience of the world shows that, as long as new land of the necessary quality is available, increased food will be obtained less by increased skill and expenditure on old land than by taking up new land. Our map has shown for several decades well over a hundred million acres in the British provinces of India classified as cultivable waste. Why is not new land coming into cultivation? I cannot give a complete answer. No such process can be observed in steady operation on a scale sufficient to raise the per capita area of cultivation to a level which will meet our food requirements. Some recent settlements in this province show an increase in cultivation of only one to three per cent in thirty years, while in others the area is stationary. For a number of reasons the area of cultivable waste gives an unreal conception of our resources. Much of the land thus classified includes areas physically capable of being employed for crops only when our need is so extreme that considerations of cost of utilization are relatively secondary. Fifty per cent we know is situated in Burma and Assam, out of the sphere of action of our chief agricultural races. A great deal is in tarai tracts where health reasons prevent extensive settlement. Land is coming under the plough, to some extent, in the villages of the Sarda Canal area in these provinces, and will do so elsewhere as irrigation schemes mature, but in India, as in other parts of the world, new land of the necessary quality for food crops is no longer easy to find.

This brings me to the first part of my argument--the necessity of increasing the acre yield of land now under the plough if an ample supply of food and the home-grown necessaries of life is to be assured to the Indian worker, and his standard of living raised above subsistence level. It is a difficult problem but it is not insoluble.

When I considered this matter some months ago, I asked myself three questions:--

(1) What factors are in our favour, and what are against us, when we begin to intensify our cultivation?

(2) Will the knowledge and experience of other countries help to accelerate our progress? What new knowledge do we need?

(3) What is the quantitative measure of the result we may expect?

I propose to give you the answers that suggested themselves to me, based on conditions in these provinces where my experience has been gained.

We have in our favour two things. In the first place, soil that is easy to manage and quickly responds to treatment, and, secondly, agricultural workers attached to their calling and possessing a strongly developed land sense which, by some curious twist in our make-up, can only be acquired in childhood. We shall not come up against a shortage of agricultural workers of the kind that is hindering development in Australia and Canada. In these countries, a high degree of skill has to be directed to economy of labour by the use of machinery and labour-saving devices. In India, our efforts will have to be devoted to economizing land. We are better placed than most countries as regards the primary essential for increasing production per unit of land, namely, man-power. You may ask me, 'What is delaying our progress with two such assets?' This opens up a wide sociological study. I believe ignorance and a larger share of ill-health than should fall to the lot of an average being play a part. The stimulus required seems to be education of a rural type. I cannot, however, pursue this issue, and return to my agricultural text.

We have to contend against difficult weather conditions and short growing seasons requiring early maturing and specialized varieties of crops. The Howards, in The Development of Indian Agriculture, describe graphically the effect of the monsoon on the soil and on the people. It is indeed the dominant factor in rural India.

We shall always at intervals experience years of short rainfall and this fact gives additional force to my argument for increasing the acre yield in favourable seasons by improved soil management if we are to avoid starvation. Much has been done to intensify yields without any commensurate increase of labour on soil improvement by the introduction of more heavily cropping varieties. I need only quote as examples wheat and cotton in the Punjab and wheat and sugar-cane in the United Provinces, which are adding crores to the cultivators' income. Indian conditions, however, test the skill of the plant breeder very severely and further steps in improvement in this direction are not going to be easily won.

I now pass on to that part of my subject which has greater interest for a scientific audience than some of the stubborn facts I have placed before you. I mean the consideration of some aspects of recent work on soil improvement and the lines on which enquiry may be directed in India.

Since Boussingault introduced the method of exact field experiment in 1834, research on the soil and the conditions of crop growth has been continuous in Europe and America. The methods of approach have become more exact with each advance in pure science. We, therefore, start our work on soil improvement in India with tools ready made. Investigations carried out in other countries have given us the principles involved and often the technique of methods of research. Our work for the moment is to apply them to conditions where soil processes differ widely both in intensity and time of occurrence, from those of temperate climates. I have been impressed by the desirability of applying to our problems a conception developed in recent years by the Cambridge and Rothamsted workers, which has given a new and wider significance to the field experiment. The final yield gives us no indication of what happens during the plant's life or how it responds to factors operating at successive stages of growth. The modern method makes quantitative observations of crops throughout the period of growth and examines the results by statistical methods This is nothing more than reducing to exact measurement and scientific treatment the observations which every practical farmer makes but does not formulate. The advantage is obvious. Information covering a wider range than the old type of field experiment can be obtained in a few years, instead of taking generations. You will remember that Lawes and Gilbert waited twenty years before discussing the results of their experiments. The field experiment lasting twenty or more years no longer fulfils our requirements. We want results in a reasonable time, accompanied by proof of their reliability, which will tell us not only the final yield but how that yield is obtained.

This leads up to another conception, namely, the critical periods of crops which will repay closer quantitative study in a country characterized by singularly short growing periods and rapidly changing conditions. By critical period, I mean the relatively short interval during which the plant reaches the maximum sensibility to a given factor and during which the intensity of that factor will have the greatest effect on yield. These periods seem to be associated with some phase of growth in which the plant is undergoing modifications demanding the rapid formation and movement of food material. Italian workers have found that the twenty days before the crop comes into ear constitutes an important critical period for wheat in relation to humidity and soil moisture. If during this period these factors are in defect of the minimum needed for the normal development of the plant, the crop will be small even if there is abundance throughout the rest of the vegetative period.

Our observations at Shahjahanpur indicate that two periods in the growth of sugar-cane have special significance: (1) May and early June when the tillers and root system are developing; and (2) August and September when the main storage of sugar takes place. A check received at either of these periods permanently reduces the yield. The acre yield of sugar is positively and closely correlated with the amount of nitrate nitrogen in the soil during the first period, and with soil moisture and humidity in the second period.

Food crops pre-eminently demand combined nitrogen. You will remember how Sir William Crookes startled the world thirty years ago by the statement that the wheat-eating races were in deadly peril of starvation owing to the rapid exhaustion of soil nitrogen. The age in which he lived had become accustomed to abundant supplies of cheap food from the great plains of the American Continent. Fertility accumulated since the glacial period by luxuriant plant growth and bacterial activity suddenly became available for exploitation, and was plundered at an appalling rate by rough and ready methods of cultivation. Nitrogen was disappearing from the soil out of all proportion to the amount recovered in the crop. The extraordinary fertility of some of these new regions is shown by the data recorded by Shutt, an acre of soil to a depth of one foot containing from 20,000 to 25,000 lb. Of nitrogen in an acre foot of soil in these provinces, which lies between the limits of 1,000 and 3,000 lb, I shall refer to this again shortly.

Crookes was almost the first to realize that there was a limit to cheap production from new land, but his forecast was too gloomy. He visualized the exhaustion of the chief granary of the western world within a generation or two. In some important respects he misapprehended the problem.

He did not know as we know now that other agencies step in and stop the plunder of the soil before it has gone too far. It is only under improper methods of cropping and cultivation that permanent soil deterioration is a real and dangerous phenomenon. Land properly handled does not become exhausted. Much of the land of Europe has been cultivated since the days of the Romans or even earlier. It is, if anything, more fertile than ever. In India, we have in existence a method of farming which has maintained for ten centuries at least a perfect balance between the nitrogen requirements of the crops we harvest and the processes which recuperate fertility.

When we examine the facts, we must put the Northern Indian cultivator down as the most economical farmer in the world as far as the utilization of the potent element of fertility--nitrogen--goes. In this respect he is more skilful than his Canadian brother. He cannot take a heavy overdraft of nitrogen from the soil. He has only the small current account provided by the few pounds annually added by nature, yet he raises a crop of wheat on irrigated land in the United Provinces that is not far removed from the Canadian average. He does more with a little nitrogen than any farmer I ever heard of. We need not concern ourselves with soil deterioration in these provinces. The present standard of fertility can be maintained indefinitely. This is not my text. Production must be raised if we are to live in reasonable security and comfort.

In one respect Crookes was right. He foresaw that the intensification of production required more combined nitrogen than the limited supplies furnished by the distillation of coal and the nitrate deposits, to counterbalance the colossal wastage which civilization and urban life bring about. The fixation of atmospheric nitrogen was, as he put it, vital to the progress of civilized humanity. This problem has been solved in the last ten years and is one of the remarkable achievements of applied science. It could have been solved sooner if money had been forthcoming for long-range research, but it took the War to bring us to our senses. Thirty years ago, the fixation of 29.4 grams of a mixture of nitrogen and oxygen at the expenditure of one horse-power was recorded as a scientific achievement. In 1928-29 the estimated production of nitrogen compounds by synthetic processes was equivalent to 1.3 million metric tons of pure nitrogen, or over 6 million long tons of sulphate of ammonia, which can be sold at prices low in comparison with the prices of agricultural produce. We are entering on an era of nitrogen plenty which is bound to react favourably on the world's food production. One of our problems is to find out how we can make use of this discovery in India. The probability is that the full benefit of fertilizers will be realized only on land reasonably supplied with organic matter.

I may be allowed here to sound a note of warning. Great as are the possibilities offered by synthetic nitrogen compounds there is danger in adjusting our standards of living to increased production based entirely on imported fertilizers. They may be cut off suddenly by international disturbances. The War is too near an experience and the promise of universal peace too uncertain to ignore this side of the question altogether. It will be but a wise precaution to establish their manufacture in India when the correct way of using them has been worked out their value demonstrated, and a demand created.

106

Our problem is more complex than the simple addition of nitrogen compounds to the soil. We have to face under peculiar conditions of climate the question of controlling moisture, organic matter, and air supply in the soil, of regulating the supplies of nitrogen so that it may be available in the right form and quantity when the plant most needs it, so that none may be wasted, and to make use to the utmost of those processes by which nature supplies nitrogen free of charge. These problems centre around the changes which organic material undergoes in the soil and the nitrogen transformations which accompany them.

We have two methods of soil improvement possessing enormous potentialities for increasing crop production and so simple in operation that they can be used by everybody:--

(1) the preparation of quick-acting manures from waste organic material;

(2) the use of green manure crops.

I do not propose to discuss recent work on the first method. The practical details have been worked out thoroughly by the Howards at Indore, and by Fowler, Richards, and their co-workers at Cawnpore. A paper on this subject is going to be placed before you by Dr. Fowler. I will not anticipate what he is going to say beyond remarking that the results which he has allowed me to examine place in our hands a method of the greatest value for increasing the out-turn of rabi crops which require in this province a quicker acting manure than that provided by turning in a green crop.

We have been working for some years at Shahjahanpur on the utilization of green-manure for sugar-cane. We have ploughed in on an average of three years' observations,218 maunds per acre of sanai (Crotalaria juncea) which adds 50 maunds of dry organic material and 75 lb. of nitrogen to each acre. We have succeeded in raising crops to 850 maunds per acre without the addition of any fertilizing agent other than the sanai produced by the land itself. . . . The practical result is worth Rs. 90 per acre. Our problem is to find out the conditions of cultivation necessary to decompose sanai in such a way that: (1) well-aerated soil containing sufficient organic matter to prevent rapid drying out is ready for the crop in March; and (2) the nitrogen exchanges are such that this element is protected from loss until it is wanted, and is then present in a form which can be rapidly mineralised for the use of the young crop.

Our method of soil treatment is to bring about the early stages of decomposition in the presence of ample moisture. The rainfall after the sanai is ploughed in is carefully watched. If it is less than five inches in the first fortnight of September the fields are irrigated. In this way we secure in most of our soils an abundant fungal growth as the land slowly dries. We prevent large accumulations of nitrates in the autumn, which may be lost before the sugar-cane is sown, and concentrate the nitrogen in easily decomposable organic form in mycelial and microbial tissue, until it is wanted in mineral form in the spring.

Throughout the experiments we have made estimates of nitrate. . . . The accumulation of nitrate reaches its maximum in May and June just before the first heavy rain. At this time the crop is about one-third grown. We have not observed any subsequent large formation of nitrate up to the completion of growth in October. The final yields are in proportion to the mineral nitrogen present in the first period and this suggests at once the importance of available nitrogen in the early stages of the growth of sugar-cane. This view is by no means a new one. It has recently been developed by Gregory at South Kensington and Rothamsted, who found that barley absorbed 90 per cent of its total nitrogen when it had made about one-third of its growth. If it is substantiated by further work and found to apply to all crops it gives a clue to several improvements in soil management.

In our studies in connexion with the intensification of sugar-cane cultivation we have been influenced by American investigations and methods, more specially those of the workers led by Waksman, who have studied the decomposition of cellulose and dead organic material in the soil. They have shown that the structure of the carbonaceous energy material in the soil largely determines the type of decomposition and the nitrogen transformations. If moisture and temperature conditions are favourable, the decomposition of cellulosic energy material, the chief constituent of green-manure, is mainly accomplished by fungous activity resulting in the formation of large quantities of mycelial tissue and the removal of nitrogen temporarily from the reach of higher plants. The synthesized material is later decomposed by other micro-organisms forming mineral nitrogen and humic material, and a definite period of time is required to complete these changes. A large volume of work has been published in the last five years. It explains much that was obscure regarding the utilization of green-manure in India, particularly the time factor to which Howard drew attention many years ago.

I now approach the last and most difficult part of my task, to estimate the increased production we may look for by the application of scientific methods to our agriculture. What I am going to say will be more readily understood if I give the production of wheat in a few countries for the crop sown in 1926, which was, on the whole, a good year throughout the world. It is as follows:--

United Provinces: Irrigated, 12.2 mds. per acre.
United Provinces:Unirrigated, 8 .2 mds. per acre
Canada, 13.2 mds. per acre
U.S.A., 10.7 mds. per acre
France, 13.0 mds. per acre
Germany, 17. 5 mds. per acre
Great Britain, 22.5 mds. per acre
Belgium,. 26.3 mds. per acre

A glance at these figures shows what an immense potential increase of production is open in many countries, especially in America and India. The physical possibility or perhaps even the limit of production in the United Provinces is shown by the yield obtained at the Shahjahanpur Research Station. In 1926 it was 28.8 maunds per acre. In the last eleven years, including two in which the wheat crop was a partial failure, 243 acres have yielded 5,945 maunds or 24.4 mounds per acre. Soil and climate do not impose a serious restriction on production. We cannot, however, take one striking instance of large yields achieved on a small acreage under favourable conditions as the basis of an estimate of the future production of the country as a whole. The actual level in any country is bound to be behind the ideal, no matter how well developed educational and propaganda machinery may be.

It is safer, if such a course be possible, to consider average results obtained in countries which have been compelled to employ intensive methods, but we have no adequate basis of comparison with our conditions. There is no example of a tropical or semi-tropical country in which scientific have been applied over a wide area by independent and unsupervised workers.

Sugar-cane cultivation in Java is often quoted as an example of what can be done. It illustrates the combined effect of strictly supervised labour and scientific methods on about one million acres of land, carried out with the object of gaining the highest possible interest on Dutch capital. It does not illustrate what we are aiming at in India-- agricultural improvement initiated and carried through by the people themselves, as the result of education and uplift, on 300 million acres.

Let us examine the course of events in Europe and America and learn what we can from them.

In medieval England the yield of wheat was seven maunds per acre. When the consolidation of holdings was completed by the enclosures in about the last quarter of the eighteenth century the yield rose to fourteen maunds per acre. It remained at this level until 1840 when a further advance was made possible by the use of better methods and the introduction of nitrogen fertilizers. By 1870 the yield had risen to twenty maunds per acre.

In America low yields and a growing industrial population are causing uneasiness. By studying agricultural conditions in other countries the conclusion has been reached that forty-seven per cent represents a possible all-round increase of production on the present cropped area. Experts do not agree as to the probable increase in the next few decades. This is placed between the limits of ten and thirty per cent. These figures are based on considerations of labour. This, as I have said, scarcely enters into our problem in India. We have more people employed in agriculture per unit of cultivated land than any other country, with the possible exception of China and Japan.

The improvement of sugar-cane cultivation extends over 2,810,000 acres in eighteen districts in the United Provinces and gives some indication of the possible course of events.

The yield of the unimproved crop in a year of average character is 350 maunds per acre. We pass through four definite stages of improvement:--

THE WASTE PRODUCTS OF AGRICULTURE

(1) Better cultivation of the old varieties, yielding 450 maunds per acre.

(2) The introduction of heavier cropping varieties accompanied by a further improvement in cultivation, yielding 600 maunds per acre.

(3) The introduction of some fertilizing agent, such as green-manure, yielding 800 maunds per acre.

(4) The intensive cultivation of heavy cropping varieties, yielding 1,000 maunds per acre.

The increase over the normal production is 28, 71, 128 and 185 per cent. The analysis of the returns is helpful in connexion with our problem. In the more important sugar producing districts seventy per cent of the sugar-cane area is planted with heavier yielding varieties. In some thirty per cent, and in a few only two per cent.

2,810,000 acres is almost exactly 33 per cent of the total sugar-cane area in the 18 districts for which special returns are made; on this area the yield has been slightly more than doubled so that there is an all-round increase in production of 33 per cent. This has taken 17 years to accomplish and brings the cultivator in 311 lakhs of rupees extra a year. I believe if such simple modifications of practice as the use of green-manure crops and composts made from waste material, were applied to all our arable land, production would be more than doubled; but this means that every cultivator would be conducting his agricultural operations in a scientific manner--a state of affairs not yet reached in any country. The point is that it is not to be expected. We must allow for the inertia which will retard the general adoption of improvements in so large a country as India. After giving due weight to this and taking into consideration the abundance of our labour re-sources and the extraordinary response of our soil to better treatment, it is reasonable to believe that within the next two or three decades we may increase the all-round out-turn of our cropped land by 30 per cent in normal seasons. But I assume that much more money will be spent on scientific research and extension work in villages than is now spent.

I hope I have said enough to show that soil improvement in India is worth an effort. It requires generous expenditure from the na-tional exchequer, and there is no better investment for it gives, to use the words of Huxley, an immediate return of those things which the most sordidly practical man admits to have value. We are working in times well suited for agricultural development. Indifference is giving way. There is a stir throughout the countryside. We can call the movement what we like, but the plain fact is that men are no longer satisfied with a life which provides only hard work and barely enough to eat. Many things are being suggested, but they deal more often than not with preliminaries to social well-being and leave untouched the vital problem of producing more food. In the end the scientific worker will come to the rescue, and the solution will be reached through the experiment station.

C. NITROGEN TRANSFORMATION IN THE DECOMPOSITION OF NATURAL ORGANIC MATERIALS AT DIFFERENT STAGES OF GROWTH

S. A. WAKSMAN and F. G. TENNEY, New Jersey Agricultural Experiment Station, U.S.A. (Proceedings and Papers of the first International Congress of Soil Science, Washington, D C, 1927, p. 209.)

To be able to understand the reasons for the rapidity of liberation of nitrogen from the decomposition of plants at different stages of growth, we must know the composition of the plant at these various stages and the nature of decomposition of the various plant constituents. Although the plant continues to assimilate nutrients, including nitrogen, until maturity, the percentage of nitrogen in the plant reaches a maximum at an early stage, then gradually diminishes, reaching a minimum at maturity or a little before maturity. This is true not only of nitrogen but also of certain other elements.

Plant materials decompose more rapidly and the nitrogen is liberated more readily (in the form of ammonia) at an early stage of growth and less so when the plant is matured. Two causes are to be considered here: (1) the rapidity of decomposition of the various plant constituents; (2) the relation of the nitrogen to the carbon content of the plant tissues.

At an early stage of growth, the plant is rich in water-soluble constituents, in protein and is low in lignins. When the plant approaches maturity, the amount of the first diminishes and of the second increases. The water-soluble constituents, the proteins and even the pentosans and celluloses decompose very rapidly provided sufficient nitrogen and minerals are available for the micro-organisms. The lignins do not decompose at all in a brief period of time of one or two months. More so, their presence has even an injurious effect upon the decomposition of the celluloses with which they are combined chemically or physically. The larger the lignin content of the plant the slower does the plant decompose even when there is present sufficient nitrogen and minerals.

It has been shown repeatedly that the organisms (fungi and bacteria) decomposing the celluloses and pentosans require a very definite amount of nitrogen for the synthesis of their protoplasm. Since the cell substance of living and dead protoplasm always contain a definite, although varying, amount of nitrogen and since there is a more or less definite ratio between the amount of cellulose decomposed and cell substance synthesized, depending of course upon the nature of the organisms and environmental conditions, the ratio between the cellulose decomposed and nitrogen required by the organisms is also definite. This nitrogen is transformed from an inorganic into an organic form. Of course in normal soil, in the presence of the complex cell population, the cell substance soon decomposes, a part of the nitrogen is again liberated as ammonia and a part remains in the soil and is resistant to rapid decomposition. The amount of nitrogen which becomes available in the soil is a balance between the nitrogen liberated from the decomposition of the plant materials and that absorbed by the micro-organisms which decompose the non-nitrogenous and nitrogenous constituents. The younger the plant, the higher is its nitrogen content and the more rapidly does it decompose, therefore the greater is the amount of nitrogen that becomes available. The lower the nitrogen content of the plant the less of it is liberated and the more of it is assimilated by micro-organisms.

These phenomena can be brought out most clearly when the same plant is examined at different stages of growth. The rye plant was selected for this purpose. The seeds were planted in the fall. The samples taken on April 28th (I), May 17th (II), June 2nd (III), and June 30th (IV). In the third sampling the plants were divided into (a) heads, (b) stems and leaves. The fourth sample was divided into (a) heads, (b) stems and leaves, (c) roots. The plants were analysed and the rapidity of their decomposition determined, using sand or soil as a medium and 2 g. of the organic matter. In the case of sand some inorganic nitrogen and minerals were added and a soil suspension used for inoculation. The evolution of carbon dioxide and accumulation of ammonia and nitrate nitrogen was used as an index of decomposition.

Tables I shows the composition of the plant and the amount of nitrogen made available after 26 days of decomposition.

TABLE I COMPOSITION OF RYE STRAW AT DIFFERENT STAGES OF GROWTH ON A DRY BASIS

No. of sample	Moisture content at time of harvest	Ash	Nitrogen	Cold water soluable fraction	Pento-sans	Cellu-lose	Lig-nin
I	80.0	7.3	2.39	32.6	15.9	17.2	9.9
II	78.8	5.7	1.76	22.0	20.5	26.1	13.5
IIIa	57.4	4.9	1.01	18.2	22.7	30.6	19.0
IIIb	60.2	5.9	2.20	20.3	22.7	20.1	16.0
IVa	15.0	3.2	1.22	4.7	11.9	4.6	13.4
IVb	15.0	3.7	0.22	9.5	21.7	34.6	18.8
IVc	?	?	0.55	4.7	26.6	37.7	21.0

When a plant material contains about 1.7 per cent nitrogen, as in the rye of the second sampling, there seems to be sufficient nitrogen for the growth of micro-organisms which decompose this material more or less completely. When the plant material contains less than 1.7 per cent of nitrogen, as in the case of the stems and leaves of the third preparation, additional nitrogen will be required, before the organic matter is completely decomposed (speaking, of course, relatively, since if a long enough period of time is allowed for the decomposition, less additional nitrogen will be needed). If the organic material contains more than I.7 per cent nitrogen, as in the case of the plants in the first planting and the heads of the third sampling, a part of the nitrogen will be liberated as ammonia, in the decomposition processes. . . . . The decomposition of 10g. dry portions of the second sampling and 20g. dry portions of the stems and leaves of the fourth sampling was studied separately in a sand medium containing available nitrogen and minerals. Only the data for the organic matter portion, insoluble in ether and water, are reported. The results show that the pentosans and celluloses are rapidly decomposed, while the lignins are affected only to a very inconsiderable extent. The nitrogen figures are of direct interest here. Just about as much insoluble protein was left in the first as in the second experiment: in the first the protein is considerably reduced, in the second increased. This tends to explain the activities of the micro-organisms in the soil. The results show that since there is a very definite ratio between the energy and nitrogen consumption of the microorganisms decomposing the organic matter, it is easy to calculate, given a certain amount of plant material and knowing its nitrogen content, whether nitrogen will be liberated in an available form or additional nitrogen will be required within a given period of time. Calculations can also be made as to how much of this nitrogen is required for the decomposition of the plant material and how long it may take before the nitrogen is again made available.

D. AN EXPERIMENT IN THE MANAGEMENT OF INDIAN LABOUR

By ALBERT HOWARD, C.I.E. (International Labour Review, Geneva, 18, 1931, p. 636.)

One of the outstanding problems of the present phase of colonial development in Asia and Africa is that of the best and most scientific methods for the organization of work in large-scale agricultural undertakings. The author of this short article, who is a well-known authority on tropical agriculture and has for thirty years contributed to the scientific improvement of agriculture in the East as Imperial Economic Botanist at the Government of India Research Station at Pasa, at Quetta, and latterly in the State of Indore, describes a small-scale experiment from which many lessons may perhaps be drawn. The experiment has been tried in the State of Indore under the stimulus of having to obtain an adequate labour force to carry on the work of an agricultural experimental station in competition with the rival attractions exercised by work in neighbouring factories. No doubt the conditions are not entirely on all fours with those of plantations carried on under competitive conditions, but they are sufficiently similar to give the experiment a living and practical interest. As the author points out, the financial basis is provided mainly by the cotton industry in India and by the Indian States members of the Institute of Plant Industry, without any call for assistance by the Government of India or by Provincial Governments. As the article shows, the best results have been obtained under a scheme which provides for a six to seven and a half hour working day, paid leave, medical attendance, good housing, and opportunity for promotion for the labour employed. [Ed. International Labour Review]

THE WASTE PRODUCTS OF AGRICULTURE

The foundation of the new Institute of Plant Industry at Indore in Central India in October 1924, provided the opportunity of breaking new ground in at least four directions, namely:--

(1) The best method of applying science to crop production.

(2) The general organization and finance (including audit) of an agricultural experiment station.

(3) The most effective way of getting the results taken up by the people; and

(4) The management of the labour force employed.

The present article deals with the last of these items: with the methods by which a contented and efficient body of labour can be maintained for the day to day work of an agricultural experiment station, largely devoted to the production of raw cotton.

THE INSTITUTE AND ITS ORGANIZATION

The Institute of Plant Industry at Indore is supported by an annual grant of Rs. 1,15,000 from the Indian Central Cotton Committee and by subscriptions, amounting at the moment to Rs. 47,550 a year, from twenty of the States of Central India and Rajputana. (In addition to these sources, the Institute makes use of the produce of the experimental area of 300 acres, of the royalties on its publications and of a number of miscellaneous items of income, including the fees earned for advice to individuals and bodies outside the Society.) During the financial year 1929-1930, the income from all sources was Rs. 1,79,080, the expenditure was Rs. 1,75,041. The management of the Institute is vested in a Board of Governors, seven in number, elected by the subscribers, the Director of the Institute being Secretary of the Board. It will be seen that the main source of the funds available for the payment of labour is derived from the Indian Central Cotton Committee (a statutory body representing the growers, the cotton trade and the officers engaged in research on cotton) created for implementing the Indian Cotton Cess Act of 1923: an Act which provides for the creation of a fund for the improvement and development of the growing, marketing and manufacture of raw cotton in India. This cess is

118

now levied at the rate of two annas per standard bale of 400 lb. on all cotton used in the Indian mills or exported from the country. The money available for the payment of labour at the Indore Institute is thus largely drawn from the cotton industry itself. At no period in the history of the institution has any financial assistance of any kind been asked for or obtained from the Government of India or from any of the Provincial Governments.

At the beginning, great difficulties were experienced in obtaining an efficient labour force. The Institute lies alongside the city of Indore, an important manufacturing and distributing centre with a population of 127,000. Nine large cotton mills (with 177,430 spindles, 5,224 looms, an invested capital of Rs. 1,67,97,106, and utilizing 68,000 bales of cotton a year) find work for 12,000 workers. In addition there are a number of ginning factories and cotton presses. The Institute therefore had to meet a good deal of local competition in building up its labour force. It was dearly useless attempting to recruit workers at rates below those readily obtained at the mills or in the city. Further, it soon became apparent that if the Institute was to succeed the Director would have to pay attention to the labour problem and devise means by which an efficient and contented body of men, women and children could be attracted and retained for reasonable periods.

Consideration of this problem led the Director to the conclusion that it could be solved by providing for the regular and effective payment of wages, for good housing, reasonable hours of work, with regular and sufficient periods of rest, and for suitable medical attention.

The application of these principles soon met with success. An adequate labour force has been built up, partly from men recruited locally and from the Rajputana States and partly from the wives and children of the sepoys of the Malwa Bhil Corps, the lines of which adjoin the Institute. A permanent labour force of about 118 is now employed throughout the year. In addition, a certain amount of temporary labour is employed for seasonal work.

The precise manner in which the principles above mentioned have been carried out in practice may now be described.

THE WASTE PRODUCTS OF AGRICULTURE

CONDITIONS OF LABOUR AT THE INSTITUTE

Payment of Labour

Wage rates for men on the permanent staff range from about Rs. 12 to Rs. 20 a month, while men on the temporary staff are paid 7 annas a day, women 5 annas, boys 3 to 6 annas, and girls 3 to 5 annas. After the rate of wages has been settled in each case, care is taken that: (1) the payment of wages is made at regular intervals; and (2) the wages are paid into the hands of the workers themselves and there are no illicit deductions on the part of the men who disburse the money.

Regularity of payment is a matter of very great importance in dealing with Indian labour. At Indore, workers on daily rates receive their wages twice a month--on the 18th and the 3rd, in each case at 2:30 p.m. The permanent labour is paid monthly on the third working day of the following month. To ensure that all payments are actually made according to the attendance registers all disbursements are made in the presence of two responsible members of the staff. Both of these men have to sign a statutory declaration that the payments have actually been made. The signed statements come regularly before the Director for signature, and are in due course placed before the auditors. In making payments the envelope system is used, the payee making a thumb impression in ink in the register or signing his or her name. These arrangements have been found to prevent any illicit deductions on the part of the staff. The payments are made in public; the rate of everybody's pay is known; the signing of a proper declaration in the register makes it possible to institute criminal proceedings at once for any irregularity; the Director is always available for inquiring into any complaints. That none have ever been made proves that the labourers actually receive their pay in full at regular intervals. Payment is made in coin; no attempt at payment in kind has ever been made; no shops for the sale of food exist on the estate and nothing whatever is done to influence the workers as to how they should spend their wages.

Hours of Labour

After the regular payment of wages, the hours of labour come next in importance. Indeed in India rest and wages are to a certain extent interchangeable as the workers regard any extra rest as equivalent to an increase in pay. At first, the Institute observed the ten hours' day so common in India, but this was soon given up. It was found during the hot months of April, May and June that both the labour and the cattle required more protection from the hot sun. An experiment was therefore made to reduce the hours of labour during the hot months to six daily, beginning work at sunrise and ending the day at sunset. The actual working hours of the three hot months were arranged in two shifts--four hours in the morning and two in the afternoon with a six hours' rest during the heat of the day, i.e. from 10 a.m. to 4 p.m. At the same time the work was speeded up and both labour and supervising staff were given to understand that the six hours' day in the hot months could only be enjoyed if everybody worked continuously and conscientiously.

The first result observed was a marked improvement in the health and well-being of the men and animals, probably due to the operation of two factors: the health-giving properties of the early morning air and avoidance of excessive sunlight. With the improvement in general health there was a corresponding reduction in cases requiring medical assistance. To everyone's surprise, it was found possible to speed up the work very considerably. The experiment of shortening the hours of labour was then extended to the rest of the year; working hours were reduced from ten to seven and a half.

These working periods, six hours in the hot weather and seven and a half during the rest of the year, refer to the time actually at work; an extra half hour daily is spent in travelling to and from the place of work. In no case does the working period exceed seven and a half hours except for about a week at the sowing time of the monsoon crops. During this period, both man and beast do not obtain much more than two hours off duty for food during the hours of daylight. A full ten hours' day at high pressure is then the rule, as all realize that the sowing of cotton and other crops is a race against time. As soon, however, as sowing is over, the workers enjoy an extra day's rest on full pay. The sowing of the monsoon crops is the only agricultural operation in Central India for which anything more than a seven and a half hours' day is necessary.

For three years the agricultural operations of the Institute have been conducted on the short hours system. The result has been suc-

cessful beyond all expectation. The miracle of speeding up Indian labour has been achieved and shorter working hours have led not only to
contentment but also to an increased output of work. This result has
only been achieved, however, by careful and detailed planning of the
work to be done each day. The daily work programme is drawn up by
the Assistant in charge of the farm during the previous afternoon and
submitted to the Director as a matter of routine, so that at daybreak
each day the Assistant knows at once what has to be done and no time
is lost in deciding what tasks have to be performed. The taking of the
attendance and the allocation of labour to the various tasks occupies
less than five minutes. In less than ten minutes after assembly, the
various gangs are at work in the fields. A great point is made of getting
down to the job at once. Punctuality is now the rule, and it is becoming
rare to have to deal with late arrivals.

While it is important to start work with the sun, it is equally
important to allow the labourers to reach their homes by sundown, particularly during the rains when snakes abound. Indian workers like to
reach home in daylight--a point of great importance in obtaining their
willing co-operation. Finally, it is very interesting to note that the policy of the square deal on the part of the Institute towards its labourers
as regards hours is now being answered by a natural desire on the part
of the workers to give the Institute a square deal. Less supervision is
becoming necessary; everybody realizes that a reduction in hours is
only possible if real work is done.

Leave and Holidays

The Institute is closed, except for work of extreme urgency, on
Sundays and on twelve important festivals during the year. In addition
to these sixty-four days, the permanent labourers are allowed one day's
casual leave and one day's sick leave every month provided they work
twenty-five full days during the month. In cases of injury while on
duty, they are allowed full pay up to a maximum of seven days. In the
case of temporary labour, all holidays and leave, except the extra day
allowed after the sowing of the monsoon crops, are given without pay.

THE WASTE PRODUCTS OF AGRICULTURE

Housing

As regards living accommodation, the demands of Indian labour are very modest. A roof which does not leak during the rains, a dry earthen floor, a room which can be locked up, a partially closed-in verandah which serves both as a kitchen and a store house for firewood, are all that is expected. At Indore the one-room cottages are arranged in blocks of six around an open courtyard in which four trees have been planted to provide shade. The quarters are fumigated and whitewashed once a year when any petty repairs to the roofs and brickwork are attended to.

After a storm-proof room, the next essential is a supply of good drinking water and a separate well for washing. The water used for drinking is raised by a simple wheel pump; the well is provided with a masonry coping about two feet high; no drinking vessels are allowed to be dipped into the water. In this way the risk of cholera is greatly reduced. Once a simple wheel pump is installed, the labourers and their wives never attempt to lower a bucket by means of a rope.

Provident Fund

So far no provident fund for the workers has been instituted. The existing provident fund only applies to the permanent staff of the Institute drawing Rs. 30 per month or more. Till the completest confidence between the workers and the management has been achieved, any suggestion of keeping back the pay of a labourer for a provident fund is likely to be misunderstood. It was decided to start a provident fund for the educated staff and gradually to extend its benefits to the labour force if and when a demand comes from the workers themselves.

Medical Arrangements

The workers and staff employed at the Institute obtain free medical attendance. In addition, the workers and the staff drawing less than Rs. 30 per month obtain free medicaments. The workers are examined weekly by the doctor so that any precautionary treatment or any advice can be given in good time. In cases of childbirth the services of a nurse are provided free of charge. The personality of the Sub-Assistant Surgeon dealing with Indian labour is very important. The workers deal with an unpopular man in a very effective fashion-- they never make use of his services.

Certificates and Promotion

An experimental station, like any employer of labour, needs some system by which the labour force can automatically renew its youth. The annual export of trained labour to centres at which improvements are being taken up is one of the important functions of the Institute. For these reasons, therefore, a supply of promising recruits must be arranged. To bring this about some system of promotion for proved efficiency had to be devised. At first this took the form of an annual promotion examination for the ploughmen. As they increased in efficiency and could manage and assemble their implements and also plough a straight furrow, their pay was increased by Re. I per month. This system is now being superseded by the certificate plan. All the permanent workers in the Institute are eligible for special training so that they can earn efficiency certificates for such operations as: (1) cultivation and sowing; (2) compost making and the care of the work cattle; (3) improved irrigation methods, including the cultivation of sugar-cane by the Java method; (4) the manufacture of sugar (Plate XIV). A certificate of efficiency (with suitable illustrations) signed by the Director can be awarded for proficiency in all these items. Each certificate which is awarded annually will carry with it an increase of Rs. 1 per month on the basic pay. When a member of the labour force has gained all four certificates, he will become eligible for transfer to other centres on higher pay. In this way the Institute holds out hope and places it within the power of any man to increase his starting pay in four years by about thirty per cent. It also enables an ambitious labourer to save enough money in a few years to purchase a holding and to become a cultivator. This is now taking place. Every year a few of the labourers return to their villages with their savings to take up a holding on heir own account. Others are deputed for work in the Contributing States on increased pay. The vacancies are automatically taken either by younger members of the same family or by volunteers on the waiting list of temporary workers.

CONCLUSION

It is possible that the system described in this article is only fully realizable on a farm working under model conditions. Nevertheless, there are a certain number of elements in this experiment which the writer feels are of universal validity in dealing with primitive labour. From the point of view of the worker it is perhaps most essential that he should feel that he is receiving a square deal. From the point of view of the management the best results are obtained by scrupulous attention to pay, by short hours of intensive work, by proper housing and medical care, and by interesting the worker in the undertaking through giving his work an educational value.

THE END

Also from Benediction Books ...
Wandering Between Two Worlds: Essays on Faith and Art
Anita Mathias
Benediction Books, 2007
152 pages
ISBN: 0955373700

Available from www.amazon.com, www.amazon.co.uk

In these wide-ranging lyrical essays, Anita Mathias writes, in lush, lovely prose, of her naughty Catholic childhood in Jamshedpur, India; her large, eccentric family in Mangalore, a sea-coast town converted by the Portuguese in the sixteenth century; her rebellion and atheism as a teenager in her Himalayan boarding school, run by German missionary nuns, St. Mary's Convent, Nainital; and her abrupt religious conversion after which she entered Mother Teresa's convent in Calcutta as a novice. Later rich, elegant essays explore the dualities of her life as a writer, mother, and Christian in the United States-- Domesticity and Art, Writing and Prayer, and the experience of being "an alien and stranger" as an immigrant in America, sensing the need for roots.

About the Author

Anita Mathias is the author of *Wandering Between Two Worlds: Essays on Faith and Art.* She has a B.A. and M.A. in English from Somerville College, Oxford University, and an M.A. in Creative Writing from the Ohio State University, USA. Anita won a National Endowment of the Arts fellowship in Creative Nonfiction in 1997. She lives in Oxford, England with her husband, Roy, and her daughters, Zoe and Irene.

Visit Anita's website
    http://www.anitamathias.com,
and Anita's blog
    http://theoxfordchristian.blogspot.com, (Dreaming Beneath the Spires)

The Church That Had Too Much
Anita Mathias
Benediction Books, 2010
52 pages
ISBN: 9781849026567

Available from www.amazon.com, www.amazon.co.uk

The Church That Had Too Much was very well-intentioned. She
wanted to love God, she wanted to love people, but she was both
hampered by her muchness and the abundance of her posses-
sions, and beset by ambition, power struggles and snobbery.
Read about the surprising way The Church That Had Too Much
began to resolve her problems in this deceptively simple and en-
chanting fable.

About the Author

Anita Mathias is the author of *Wandering Between Two Worlds:
Essays on Faith and Art.* She has a B.A. and M.A. in English
from Somerville College, Oxford University, and an M.A. in
Creative Writing from the Ohio State University, USA. Anita
won a National Endowment of the Arts fellowship in Creative
Nonfiction in 1997. She lives in Oxford, England with her husband, Roy, and her daughters, Zoe and Irene.

Visit Anita at http://www.anitamathias.com, and on
http://theoxfordchristian.blogspot.com, her Christian blog;
http://wanderingbetweentwoworlds.blogspot.com/, her personal blog, and
http://thegoodbooksblog.blogspot.com, her literary and writing blog.

www.ingramcontent.com/pod-product-compliance
Lightning Source LLC
Chambersburg PA
CBHW022114280326
41933CB00007B/389